Developments in Atmospheric Science, 6

Physical Principles of Micro-Meteorological Measurements

Further titles in this series

1. F. VERNIANI (Editor)
Structure and Dynamics of the Upper Atmosphere

2. E.E. GOSSARD and W.H. HOOKE
Waves in the Atmosphere

3. L.P. SMITH
Methods in Agricultural Meteorology

4. O.M. ESSENWANGER
Applied Statistics in Atmospheric Science

5. G.W. PALTRIDGE and C.M.R. PLATT
Radiative Processes in Meteorology and Climatology

Developments in Atmospheric Science, 6

PHYSICAL PRINCIPLES OF MICRO-METEOROLOGICAL MEASUREMENTS

by

PETER SCHWERDTFEGER

Professor of Meteorology,
The Flinders University of South Australia,
Bedford Park, S.A., Australia

ELSEVIER SCIENTIFIC PUBLISHING COMPANY
Amsterdam - Oxford - New York 1976

ELSEVIER SCIENTIFIC PUBLISHING COMPANY
335 Jan van Galenstraat
P.O. Box 211, Amsterdam, The Netherlands

Distributors for the United States and Canada:

ELSEVIER/NORTH-HOLLAND INC.
52, Vanderbilt Avenue
New York, N.Y. 10017

ISBN: 0-444-41489-4

Printed in The Netherlands

Preface

Meteorology is increasingly becoming part of the tertiary educational scene, perhaps because of a long overdue more general realization that the atmosphere constitutes one of the most omnipresent environments whose study is basic to a broad range of investigations ranging from physics to ecology. Far too often, meteorology is regarded as being synonymous with weather forecasting or perhaps climatology, when these are but specialized, although certainly important, applications. There are several books, including texts published by various national meteorological services, which deal admirably with the principles and use of a large variety of network instruments. This book is not intended to compete with these. Rather, it has been designed to emphasise the physical basis of precise meteorological measurements, particularly those necessary in the determination of those processes relevant to the behaviour of the atmosphere near the surface of the earth.

The experiments which have been devised show a general, but not completely monotonic, trend in difficulty and should suffice as a basis for student practical activity in physical meteorology over a two-year period. Wherever feasible, suggestions for the construction of equipment are given, although the purchase of certain items will be unavoidable. Because of the varying availabilities of various manufacturers' instruments in different countries, one of the references listed at the end of this book should prove helpful in arriving at the most advantageous selection.

Unfortunately, various meteorological observations are either precluded or made difficult under certain weather conditions, so that the sequence of experiments may require flexible selection. Because of this universal problem, many of the experiments described have been contrived to be conducted indoors in a basic laboratory. However, others unavoidably require the use of a permanent outdoor meteorological area having both grass and soil surfaces as well as substantially unobstructed views of the sky, particularly to the east and west, but preferably also to the north or south in the southern or northern hemispheres respectively.

The subject matter in this book has been evolved over a twelve-year period, partly because of the total absence of any similar published material in the English language, and used by students under my instruction at both the University of Melbourne and the Flinders University of South Australia. In its initial, abridged form, it was published as *Physical Principles of Basic Meteorological Measurements* in 1971 by the Meteorology

Department of the University of Melbourne, whose Head, Dr. U. Radok, encouraged its development. An augmented version was issued by the Flinders Institute for Atmospheric and Marine Sciences in 1972. The final details of the book in its present form were completed whilst visiting Göttingen University, Germany, under the sponsorship of the Alexander von Humboldt Foundation.

March, 1976

PETER SCHWERDTFEGER

Contents

List of Tables

CHAPTER 1

Air temperature and sensible heat transfer

A potential difference across an electrical conductor causes a flow of electric current, similarly a temperature difference between two points in a thermal conductor results in a flow of heat between them. In meteorology a great many observations are concerned with the measurement of flows of energy in heat or related form. Sensible heat transfer in the air, conduction in soil, radiation from the sun and the energy absorbed in the evaporation of moisture provide a number of examples. Although various instrumental techniques have been developed for the measurement of energy fluxes such as these, all of them, even if only for calibration purposes, require the precise measurement of temperatures or temperature differences.

1.1. METHODS OF TEMPERATURE MEASUREMENT

As the measurement of temperature is often important in its own right, as well as being the basis of the determination of other quantities, a wide range of methods have been developed. Some may be usefully incorporated in devices for the automatic recording of temperatures, while others are usually found only in sensors designed for in-situ reading.

The methods fall into six main classes which can be grouped under the following phenomena:

(1) *Expansion effects.* These depend on the thermal expansion of gases, liquids and solids. Liquid thermometers, in particular mercury or alcohol in glass or metal are well known. Solid thermometers usually rely on differences in thermal expansion between dissimilar metals. By means of linkages and levers the relative motion or distortion that occurs in a compound metal sensor on temperature change can be harnessed to drive either a visual indicator over a calibrated scale or an inked pen over a moving chart. The latter is the principle behind the well-known clockwork thermograph which can operate without the provision of external power sources.

(2) *Electrical resistance effects.* In true metals an increase in temperature results in an increase in electrical resistance. Semi-conductors such as the borderline-metal germanium and the non-metal carbon decrease in resistivity with temperature, a phenomenon resulting from the increasing overlap in energy of bands of electrons as energy is supplied in the form of heat. The semi-conductors with their negative temperature coefficient of resistance

are usually far more sensitive to temperature change than metals, a fact made use of in thermistors which are made of semi-conductors such as germanium and silicon, usually in the form of a bead or a rod.

(3) *Thermoelectric effects.* When the two ends of a metal conductor are at different temperatures an electric potential difference, called the Thomson electro-motive force (e.m.f.), is created between the ends; similarly a Peltier e.m.f. is seated at the junction of two dissimilar metals. In an electrical circuit both of these e.m.f.'s combine to give what is known as the Seebeck e.m.f. Because these e.m.f.'s depend on the temperature, the thermocouple, consisting of a circuit incorporating two junctions of two dissimilar metals, provides a means for measuring the temperature difference between the two junctions.

(4) *Radiative effects.* The temperature of identifiable emitters of radiation in bands strongly related to that temperature can be determined by suitable radiation detectors operating in the infra-red region of wavelengths. These devices are particularly suited to the techniques of remote sensing of temperatures and are used in both satellite- and aircraft-based measurements.

(5) *Acoustic effects.* The fact that the velocity of sound in dry air is only a function of the temperature is made use of in sonic thermometers. Such sensors have an almost instantaneous response in that this is dependent only on the transit time of the sonic pulse through the air sample, whose temperature, if non-uniform is effectively found as an integrated space mean.

(6) *Combined effects.* An example is provided by the electrical oscillations of suitably cut quartz crystals. As the temperature changes, the expansion or contraction of the crystal results in a variation in the frequency of oscillation. Such sensors are now the basis of extremely precise and sensitive thermometers.

1.2. SOURCES OF ERROR IN TEMPERATURE MEASUREMENT

The principal cause of error in the measurement of the temperature of any substance is the lack of similarity in the physical properties of the sensor material and the substance being investigated. In all cases, the thermal inertia of the sensor results in some delay when it is first introduced into a new substance or environment. This also implies that if the environmental temperature being observed undergoes some fluctuation, the capacity of a sensor to follow these is determined by both the fluctuation frequency and the sensor's thermal mass. The thermal inertia depends both on the thermometer sensor and on the environmental medium being sensed. For example, a mercury thermometer may require over a minute to come to an equilibrium temperature in still air but gives a stable water temperature reading within a few seconds.

While thermal inertia is determined by the relative values of the specific thermal capacity and the effective thermal conductivity of the sensor and the sensed medium, lack of similarity in optical and radiational properties generally results in serious problems for the meteorologist. Because the rate of heat transfer between the sensor and its environment may not be sufficient to dissipate the flux of radiation absorbed by the sensor under many conditions, great care has to be taken to avoid anomalous readings. This problem arises when the temperature of any transparent or translucent material is being measured including ice, snow, water and air. The seriousness of the radiation-induced heating on the sensor temperature increases as the thermal coupling with the environmental medium decreases, so that in a vacuum, the intensity distribution of radiation alone determines the temperature of a thermometer sensor.

1.3. SENSOR THERMAL INERTIA

Let the temperature of the sensor be T and that of the surroundings, air for example, be T_a. Should the two differ then heat will flow across the sensor boundary. When heat flows from the air to the thermometer the flux of heat is taken as being positive:

$$H = \omega(T_a - T) \qquad [1.1]$$

ω is a heat transfer coefficient depending on the shape of the boundary across which transfer is taking place and on the air velocity, or, more precisely, on the effective thermal conductivity of the energy transferring medium surrounding the sensor. The effective conductivity of still and moving air can vary by orders of magnitude.

In the absence of other modes of heat exchange such as radiation, the sensible heat flux in the surrounding air, H, is solely responsible for a gradual change in temperature of the thermometer element. Although this is rarely the case, H can, for the moment, be considered as effectively a sensible heating of the air and not radiative exchange with surrounding objects. If temperature differences are not too great, a linear relationship as shown in equation [1.1] can be considered to apply. However, the coefficient ω would not be entirely independent of the temperature.

Assuming the temperature sensor to have good conductivity then the heat loss may be considered as a flow across the boundary, and is given by:

$$H = \frac{c\rho V}{A}\frac{dT}{dt} \qquad [1.2]$$

where t is the time, A is the surface area of the sensor, V its volume, ρ its density and c the specific heat of the sensor substance. It thus follows from [1.1] and [1.2] that:

$$\frac{dT}{dt} = \frac{\omega A}{c\rho V}(T_a - T) \qquad [1.3]$$

which expression may be recognised as a form of Newton's Law of Cooling. Integration of equation [1.3] leads to the following expression:

$$(T - T_0) = (T_a - T_0)(1 - e^{-(\omega A/c\rho V)t}) \qquad [1.4]$$

or:

$$(T - T_a) = (T_0 - T_a)\,e^{-(\omega A/c\rho V)t} \qquad [1.5]$$

where T_0 is the temperature of the element at time $t = 0$. The quantity $c\rho V/\omega A$ can be written as τ, the time constant. The speed with which a thermometer element follows the temperature of its surroundings depends only on τ; for a fast response and correspondingly good resolution of rapid temperature changes τ should be as small as possible. This can be achieved by choosing sensors having a small specific heat, small volume to surface ratio and of course to be of a material having a high thermal conductivity. Equations [1.2], [1.3] and [1.4] are not strictly valid for sensor substances of poor conductivity. In fact, even for thermo-elements of high conductivity they are strictly applicable only to those with at least one small dimension, for which it can be assumed that no lagging is caused by heat transfer within the sensor material itself, so that the temperature indicated by the sensor approximates that of its own surface.

Thermometers designed to reach temperature equilibrium with the surroundings quickly are, when of the liquid-in-glass or -metal type, equipped with elongated bulbs of large surface area and small volume. On the other hand, certain thermometers are designed for delayed reading after removal from the site where the temperature has been measured such as soil and snow thermometers. These are given large bulbs of high total thermal capacity and low surface area to volume ratio, characteristics which may be augmented by means of surrounding the sensor with a larger volume of wax which also acts as an insulator when such thermometers are removed from the environment they are monitoring for reading.

In certain circumstances, the technique of forced ventilation, which is often used on larger thermometers to decrease the response time by increasing the magnitude of the exchange coefficient ω, is not applicable, as the draught may destroy the effect under observation. This problem often arises in the study of only slowly moving boundary layers of air near to surfaces such as the ground and in the vicinity of plant environments. In these cases very small thermocouples or resistance thermometers with a surface area to volume ratio as high as possible provide a solution. A suitable shape for a resistance thermometer would be that of a long extremely thin wire.

Experiment I. Thermal inertia of a thermometer

In this experiment, the thermal inertia of several mercury-in-glass thermometers are examined and the heat transfer coefficient, ω, is determined for still air in a laboratory.

At least three thermometers whose bulbs are of simple and contrasting physical dimensions should be used. An example would be in selecting thermometers having spherical bulbs of different diameters.

The stems should be supported just above a table on simple supports which may be cut from rubber pencil erasers.

To commence the determination of thermal inertia of a thermometer its equilibrium reading under room temperature conditions is first observed, then it is heated slowly above an electric radiator so that the mercury achieves a uniform temperature and no conduction takes place along the stem.

When a suitable temperature has been reached the thermometer is placed on the rubber stem supports and the indicated temperature is read at suitable time intervals. The smaller the thermometer, the closer these time intervals should be.

Thus a series of bulb temperatures, T, as a function of time t may be obtained, with an initial reading of T_0 at $t = 0$ within a room temperature environment T_a. The latter should be checked again at the conclusion of the experiment so that due allowance can be made if a significant change has occurred.

Equation [1.5] can be written:

$$t = \tau \ln \frac{T_0 - T_a}{T - T_a} \qquad [1.6]$$

$$\therefore \tau = t \Big/ \ln \frac{T_0 - T_a}{T - T_a}$$

$$\text{i.e. } \tau = 0.434 t \Big/ \log_{10} \frac{T_0 - T_a}{T - T_a} \qquad [1.7]$$

From [1.6] it is evident that if log-linear graph paper is used the functional relationship between t and $(T_0 - T_a)/(T - T_a)$ will appear to be linear and obviate the need to calculate individual logarithmic values.

The line of best fit, which must pass through the point $t = 0$ and $T = T_0$ then allows the gradient to be determined. This is given by:

$$\tau = 0.434 \Delta t \Big/ \Delta \log_{10} \frac{T_0 - T_a}{T - T_a} \qquad [1.8]$$

where the Δ's indicate two suitably chosen intervals along the axes for which of course the actual (including the logarithmic) values must be employed.

Any marked departure from the theoretically anticipated functional relationship, if occurring at the beginning of the cooling process, can usually be traced to non-uniform heating of the thermometer. Heating by means of immersing in hot water of course introduces the spurious heat losses caused by evaporation.

Since $\tau = c\rho V/\omega A$, an attempt can be made to calculate ω for each of the thermometers used. For mercury, $c = 0.140\ \mathrm{J\,g^{-1}\,°C^{-1}}$ and $\rho = 13.55\ \mathrm{g\,cm^{-3}}$, so that with τ determined in seconds, ω may be found in units of $\mathrm{W\,cm^{-2}\,°C^{-1}}$.

A procedure similar to that described above that makes use of large bulb (approximately $10\ \mathrm{cm^3}$) alcohol-in-glass thermometers, has found application in simulating the cooling rate of the human body. This acknowledges the fact that in cold regions of the world, the air temperature alone provides an insufficient index of human comfort and that the level of ventilation (which determines ω) is an important further parameter.

Experiment II. Measurement of the heat transfer coefficient for a plane surface

The apparatus consists of aluminium disks set into insulating styrofoam blocks so that significant cooling takes place only at the upper surface. Holes drilled into the side of the aluminium disks allow the temperature at the centre to be read by means of mercury-in-glass thermometers. Because of the high thermal conductivity of the metal, the temperature might be considered as approximately uniform throughout except under conditions of extremely rapid surface cooling. In a more rigorous experimental arrangement, small thermocouples may be embedded just under the upper surface of the disks.

An electric blower is set up to direct its draught along a table and different groups of experimenters can work simultaneously observing the rate of cooling of disks, located at various distances along a line downwind from the fan. Since the air flow is not constrained by walls, the wind speed falls off with increasing distance from the blower and can be measured above each disk with a suitably small air flow meter. Alternatively, small micrometeorological anemometers can be incorporated into the apparatus. The latter form of instrumentation is essential for outdoor use of the equipment as the cup anemometers are insensitive to changes in wind direction.

The disks are heated slowly by a dry heat source and then placed in styrofoam insulating surrounds which themselves are incorporated into a plane layer of insulating material, leaving only the aluminium surfaces exposed to the mechanically produced draft of air. Adopting the procedure developed in Experiment I, the value of $\tau = c\rho V/\omega A$ and hence ω may be determined for a number of air speeds. Since the ratio of the volume to exposed surface area for a cylinder, $V/A = h$, the thickness of a cylindrical disk, only the latter need be measured. The hole drilled to accommodate the thermometer

Fig. 1.1. Apparatus for the measurement of heat transfer coefficient for a plane surface. Polyurethane blocks insulate all except the upper surfaces of solid aluminium disks whose temperature is monitored by mercury-in-glass thermometers. The foreground shows one of the disks, with thermometer inserted, removed from its insulator. Above each surface, small cup anemometers record the speed of the air stream issuing from the cardboard collimater in front of an electric fan. The case on the right houses the electro-mechanical counters registering anemometer rotations.

can be regarded as inconsequential since it is occupied by the thermal mass of the thermometer itself. The equipment is shown in Fig. 1.1.

The specific heat and density of aluminium may be taken to be $0.88\ \mathrm{J\,g^{-1}\,{}^{\circ}C^{-1}}$ and $2.7\ \mathrm{g\,cm^{-3}}$ respectively.

The functional relationship between ω for a plane surface and the air speed above can be plotted graphically and extrapolated to zero velocity. The latter value of ω might be compared to the values obtained in the first experiment.

Values of ω as found above are of practical value and constitute fundamental micro-meteorological information.

1.4. THE EFFECT OF RADIATION ON TEMPERATURE SENSORS

In general a thermometer element is exposed to radiation. Even if shielded from solar and sky radiation which is mainly in the visible wavelengths, there still remains infra-red radiation from the surroundings. A fraction of this radiation is absorbed, the remainder reflected or re-emitted. The thermometer element surface itself emits radiation to an extent which is determined by the Stefan–Boltzmann Law and the surface emissivity, ϵ:

$$F = \epsilon\sigma T^4 \qquad [1.9]$$

where σ is the Stefan–Boltzmann constant and T is the absolute temperature in °K. The accepted value for σ is 5.67×10^{-9} mW cm^{-2} °K^{-4}.

Radiation at the earth's surface can be considered to be grouped in two main ranges of wavelength. The sun which can be considered as a black body at about 6000°K is the source of short-wave, which includes ultra-violet and visible, radiation, some of which is scattered and reflected by the sky and terrestrial surfaces. Long-wave or infra-red radiation is emitted by the earth and its atmosphere which are at a far lower temperature, usually in the vicinity of 270 ± 50°K. The relative angular extent of solar and terrestrial sources as seen from the earth's surface, results in the flux of solar radiation not significantly overlapping with that of terrestrial origin. Since most radiators are not perfectly black bodies, with radiative properties varying over the spectrum, it is necessary to consider mean radiative properties of surfaces over visible and infra-red ranges separately. Thus the visible, or short-wave absorptivity is given by α and the infra-red emissivity by ϵ, both of these quantities being $\leqslant 1$. For a perfectly black body, they would be both equal to 1.

If a thermometer sensor (or in fact any body) is exposed to visible and infra-red fluxes of respectively S_i and L_i, then the net gain in heat energy by radiation is:

$$A_N = \alpha S_i + \epsilon(L_i - \sigma T^4) \qquad [1.10]$$

as can be deduced from Fig. 1.2.

The energy gain of the sensor by radiation must be counterbalanced by heat loss across the boundary layer; hence equation [1.2] requires modification to:

$$H + A_N = \frac{c\rho V}{A}\frac{dT}{dt} \qquad [1.11]$$

When the thermometer reaches equilibrium, $dT/dt = 0$, and:

$$H + A_N = 0$$

so that:

$$\omega(T - T_a) = A_N \qquad [1.12]$$

Hence from [1.10] and [1.12] we have for the difference between sensor and air temperature, noting that all temperatures must be in °K:

$$\Delta T = (T - T_a) = A_N/\omega$$

$$= \frac{1}{\omega}[\alpha S_i + \epsilon(L_i - \sigma T^4)] \qquad [1.13]$$

If the thermometer is shielded by a polished metal screen the magnitude

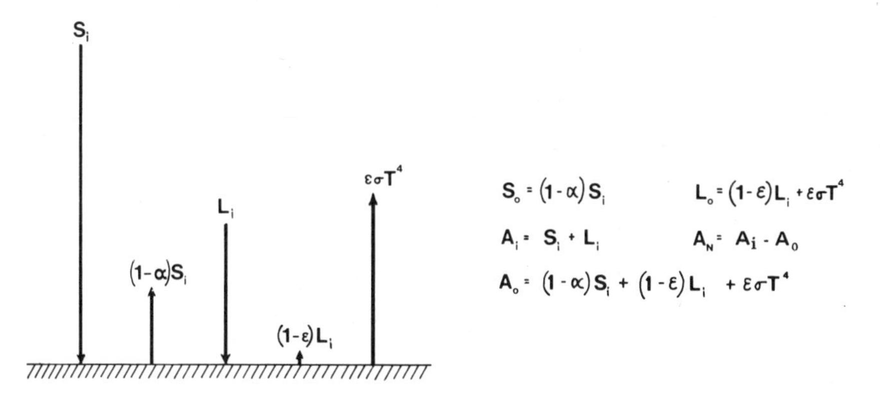

Fig. 1.2. The fluxes of radiation at a plane, opaque surface, where the subscripts "i" and "o" respectively denote inward and outward directions relative to the earth.

TABLE 1.1

Radiation fluxes $F = \sigma T^4$ (mW cm^{-2})

T(°K)	0	1	2	3	4	5	6	7	8	9
240	18.81	19.13	19.45	19.77	20.10	20.43	20.76	21.10	21.45	21.80
250	22.15	22.50	22.87	23.23	23.60	23.97	24.35	24.74	25.12	25.51
260	25.91	26.31	26.72	27.13	27.54	27.96	28.39	28.82	29.25	29.69
270	30.13	30.58	31.04	31.49	31.96	32.43	32.90	33.38	33.87	34.36
280	34.85	35.35	35.86	36.37	36.89	37.41	37.94	38.47	39.01	39.55
290	40.10	40.66	41.22	41.79	42.36	42.94	43.53	44.12	44.71	45.32
300	45.93	46.54	47.16	47.79	48.43	49.07	49.71	50.37	51.03	51.69
310	52.36	53.04	53.73	54.42	55.12	55.82	56.54	57.26	57.98	58.71
320	59.45	60.20	60.95	61.72	62.48	63.26	64.04	64.83	65.63	66.43
330	67.24	68.06	68.89	69.72	70.56	71.41	72.27	73.13	74.00	74.88

of both α and ϵ can be reduced to about 0.1. Usually, on the assumption that L_i and σT^4 are of similar magnitude it is regarded as most important to keep α particularly, as low as possible. Because α is less for a good white paint (say 0.05) than for a chromium or even average aluminium surface, Stevenson screens which are a standard form of meteorological instrument shelter, are constructed with white-painted wooden louvres which should be kept clean to reduce absorption of short-wave radiation to a minimum. The louvres are designed to permit natural ventilation of the thermometers within while achieving efficient shading. Nevertheless substantial errors in temperature readings can result under calm conditions, when ventilation is inadequate. The psychrometer designed by Assmann houses its component thermometers in highly reflective tubular metal shields and operates in a controlled draught induced by a mechanically driven fan.

Table 1.1 lists values of radiative fluxes $F = \sigma T^4$ for a range of terrestrial temperatures.

Fig. 1.3. Combinations of black- and white-surfaced shields for mercury-in-glass thermometers, whose sensing bulbs are located centrally between the circular plates by holed cork wedges.

Experiment III. Effect of radiation on shielded thermometers

Four identical thermometers are used in conjunction with four types of surrounding shield. The shields consist of two parallel disks held apart by spacers. The outside and inside surfaces of these shields are either painted white or matt black; four combinations of these surfaces are of course possible. The black surfaces can be considered to absorb all radiation incident on them and to be perfect black bodies. Being at relatively low temperatures, only infra-red or long-wave radiation will be emitted. The white surfaces might be considered to reflect all short-wave radiation but to absorb the long-wave radiation incident. Again, all re-radiation will be in the long-wave region.

Clearly, the heat loss for the thermometers occurring by convection, because of air movement through the shield space, is difficult to assess, so that in this simple form the experiment is a qualitative one only. It is interesting, however, to compare the effect of the white and black interior surfaces respectively on the thermometers they enclose.

The four thermometers within the different types of shield should be exposed to the sun and the temperatures noted over a period of about ten minutes. The greatest contrast in readings is obtained in an area sheltered from the wind. The differences of these readings with that of a similar ventilated thermometer should be measured and discussed.

An extension of this experiment is to measure the thermometer time constant, using different types of shield. The units should be sheltered from the wind, but not of course from radiation for these measurements, which should verify that the thermal inertia of a thermometer sensor is dependent upon its own physical properties and the motion of the environmental air only.

Fig. 1.3 shows a photograph of readily constructed equipment, with corks being used to hold the mercury-in-glass thermometers in position. The open design of the shields will result in differences of the four thermometers' temperatures being reduced under the ventilating influence of winds.

1.5. ELECTRICAL RESISTANCE THERMOMETERS

Electrical devices sensitive to temperature variation may be based on changes in resistance. Resistance thermometers often consist of a suitable metallic wire whose resistance increases with temperature. However, a further branch of resistance thermometry has developed with semi-conductors, in a form known as thermistors, from *therm*ally-sensitive-res*istor*.

The resistance of a metallic conductor as a function of temperature can be written as a power series:

$$R_T = R_0(1 + \alpha T + \beta T^2 + \ldots) \quad [1.14]$$

For meteorological ranges of temperature, a linear relationship between resistance and temperature is usually sufficiently accurate for metallic resistors:

$$R_T = R_0(1 + \alpha T) \quad [1.15]$$

Table 1.2 shows the value of α for a number of conductors.

The choice of a particular metal for use in a resistance thermometer is mainly centred on its permanence since apart from specially developed alloys such as manganin, used for winding accurate temperature-stable resistors, most metals have the same order of magnitude of sensitivity to temperature change. Since platinum is not subject to oxidation or corrosion it is most often used.

Thermistors consist of suitable semi-conductors of which carbon and the borderline metals such as germanium are simple examples.

In these substances, the lattice electrons do not have the same freedom of motion as in metals, but an increase in temperature results in sufficient energy to be imparted to a sufficient number of them to become free and

TABLE 1.2

Temperature coefficient of resistance of various materials

Material	$\alpha(^{\circ}C^{-1})$
Metal	
Aluminium	0.0039
Copper	0.0039
Iron	0.0062
Platinum	0.0037
Silver	0.0038
Tungsten	0.0045
Manganin	0.000000
Non-metal	
Carbon	− 0.0005

hence cause increased conduction. Hence in these materials, resistance decreases with temperature. In fact, in semi-conductors, the magnitude of the temperature coefficient of resistors is much greater than that of good conductors. Thermistors are thus valuable for measuring small temperature differences although they may need more frequent calibration than a platinum thermometer.

Experiment IV. The dissipation of heat from a resistance thermometer

Resistance thermometers, including thermistors, find frequent application in meteorology as for example in the measurement of air temperatures and particularly their gradients. Either the latter, or temporal fluctuations of the former must be known in order to be able to determine the sensible flow of heat in the atmosphere.

The simplest reliable method of electrical resistance measurement makes use of a Wheatstone bridge network as shown in Fig. 1.4. This bridge is said to be balanced when the relationship between the values of the resistances shown in the diagram is:

$$\frac{R_1}{R_2} = \frac{R_3}{R_4} \quad \text{or} \quad R_1 = R_2\frac{R_3}{R_4} \qquad [1.16]$$

It is clear that some power will be dissipated in all of the resistance elements including the sensor of resistance R_1, which will raise the latter's temperature above that of the surrounding air being monitored. Increasing the resistance of the adjustable arms R_3 and R_4 of the bridge, reduces the current flowing through the sensor but simultaneously lowers the sensitivity of the bridge system.

Since the rate of heat loss from a heated sensor increases with ventilation,

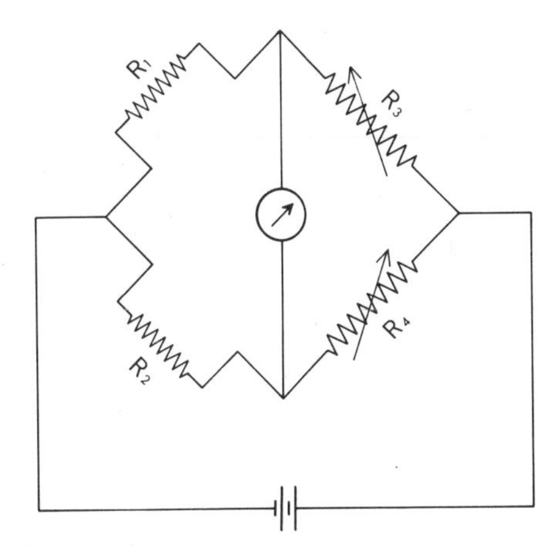

Fig. 1.4. Wheatstone bridge circuit for resistance thermometers.

one of the purposes of this experiment is to determine the minimum ventilating air speed compatible with a given accuracy requirement for air temperature indication. In order to reduce the amplitude of sensor temperature fluctuations caused by turbulent eddies in the air stream, adequate sensor thermal inertia must be ensured. If glassed bead thermistors be used for example, then for the purposes of this experiment they should be surrounded by metal sleeves in the form of cylinders approximately 1 cm in length and 0.5 cm in diameter with good thermal contact being achieved by means of using wax or petroleum jelly.

The simplest method is to arrange for the arm R_2 in Fig. 1.4 to be an element identical to R_1. Initially both of these temperature elements should be located together in a relatively high-speed air stream. This air should, of course, be at the environmental temperature. One of the sensors is now left in position to act as a reference whilst the other is sequentially moved to be ventilated by various lesser air speeds. In each case, the actual temperature difference between the two sensors should be noted for a calculable level of power dissipation within them. The accuracy of the indicated air temperature reading, as determined by the difference from that indicated by the reference sensor, should be plotted as a function of ventilation speed.

Since the sensitivity of the bridge increases with the current passing through the resistance elements, the above experiment should be repeated for a number of current levels which should be adjusted by suitably varying the supply voltage, V, only. In this way, an optimum operating current may be determined for known conditions of ventilation.

In setting up this experiment, it is preferable to balance the bridge when both sensors R_1 and R_2 are in good thermal contact at the environmental temperature by adjusting the value of either R_3 or R_4. The system should also be calibrated to indicate the difference between the temperatures of the

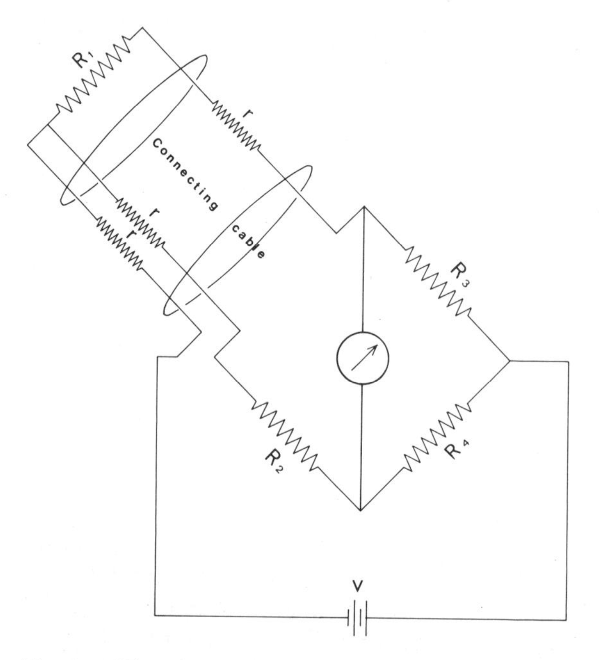

Fig. 1.5. Wheatstone bridge circuit for remotely connected resistance thermometers.

two sensors. This can be done by immersing the sensing elements in separate, well-insulated volumes of stirred water whose temperatures must not be too far removed from the environmental air temperature because of the non-linearity of the bridge. The water temperatures require independent means of measurement by any accurate thermometer and care must be taken that the electrical sensors under test are electrically but not thermally insulated from the water baths.

If the rate of electrical power dissipation in a resistance thermometer is W, then the flux of internally generated heat passing across the sensor boundary whose surface area is A is given by W/A. Hence for a resistance thermometer, equation [1.12] must be modified to include this additional term, so that:

$$\omega(T - T_a) = A_N + \frac{W}{A} \qquad [1.17]$$

and as a consequence the expression for the error in air temperature measurement becomes:

$$\Delta T = \frac{1}{\omega}\left(A_N + \frac{W}{A}\right) \qquad [1.18]$$

Again it is evident that by ensuring a sufficiently high rate of ventilation and hence high value of ω, the error ΔT in the air temperature reading can be reduced to the level required.

Assuming that a sensor can be protected so that the radiation term can be neglected, as has in fact been done in Experiment I, then:

$$\Delta T \approx W/\omega A \qquad [1.19]$$

Since W, A and ΔT can be measured, the applicability of the values of ω measured as a function of wind speed in Experiment II can be ascertained. However, since the effect of varying sensor shapes requires consideration in this evaluation, a more practical approach would be to regard equation [1.19] as being the basis of an alternative and simple method for the determination of ω itself.

It is readily seen that when the circuit shown in Fig. 1.4 is balanced, the power dissipated in the sensor of resistance R_1 is given by:

$$W = R_1 \left(\frac{V}{R_1 + R_3}\right)^2 \qquad [1.20]$$

where V is the electrical potential applied to the bridge.

In most practical applications, the resistance thermometer element R_1 will be located, perhaps on a mast, remotely from the remaining elements of the bridge circuit. To avoid temperature-dependent and hence in general varying lead resistances generating serious errors in indicated temperature readings, the circuit arrangement of Fig. 1.5 should be adopted. The balance condition then becomes:

$$\frac{R_1 + r}{R_2 + r} = \frac{R_3}{R_4}$$

so that:

$$R_1 = R_2 \frac{R_3}{R_4} + r\left(\frac{R_3}{R_4} - 1\right) \qquad [1.21]$$

where the three longer leads shown are assumed to be of equal resistance r. Equation [1.21] shows that if the bridge is designed to be symmetrical so that $R_1 = R_2$ and hence $R_3 = R_4$, then the actual value of r will be immaterial.

The effect of lead resistance on the power dissipated in the sensor is covered by the right-hand side of equation [1.20] being changed to $R_1[V/(R_1 + R_3 + 2r)]^2$ from which it is evident that if $(R_1 + R_3) \gg 2r$, the influence of the leads will be negligible.

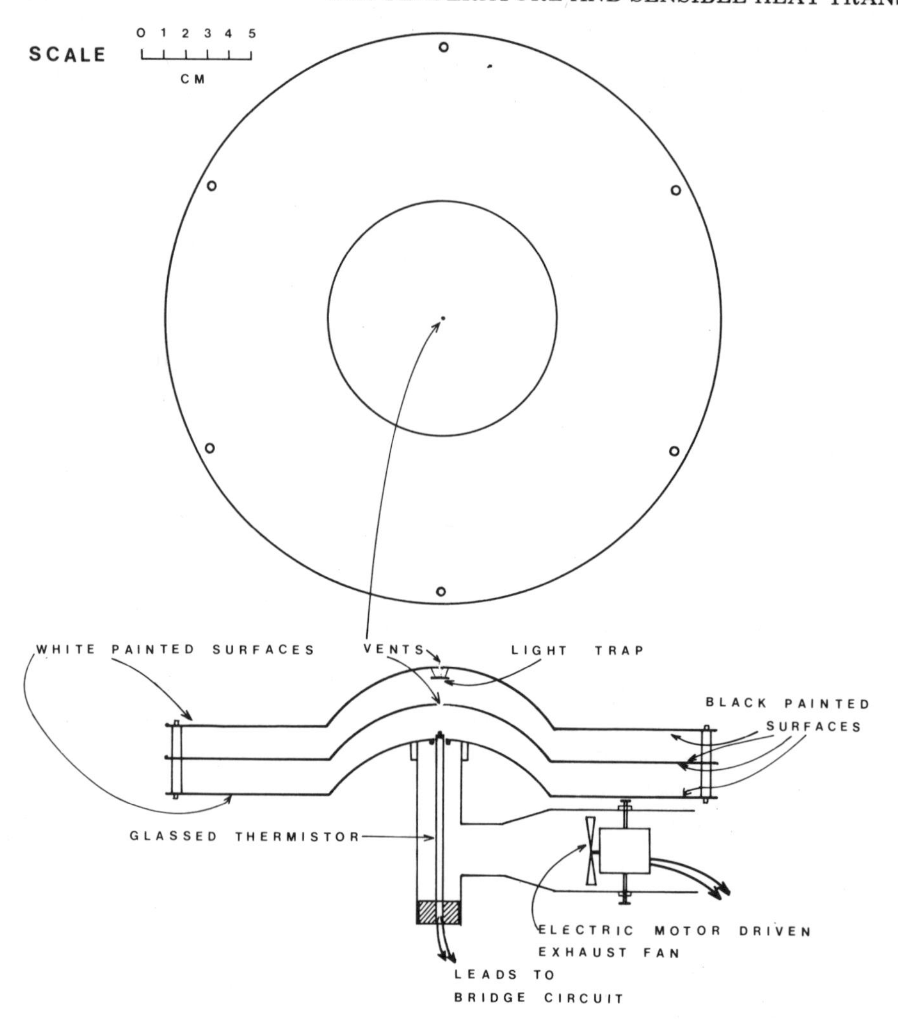

Fig. 1.6. A ventilated shield for resistance thermometers in plan and cross-section.

1.6. A VENTILATED SHIELD FOR RESISTANCE THERMOMETERS

In many micro-meteorological applications, such as the measurement of atmospheric temperature gradients in the first ten metres or so above the ground, relatively precise temperature records, with typical resolutions of differences better than $0.05\,^{\circ}\mathrm{C\,m^{-1}}$, may be required. Thus the combined effects of incident radiation and internal power dissipation necessitate careful shielding of resistance thermometer elements in particular. A successful design of mounting is shown in Fig. 1.6 and is suitable for appropriately dimensioned sensors intended for monitoring mean air temperatures over periods exceeding one minute.

The three parallel surfaces may be formed by spinning aluminium disks but these shapes could also be equally well beaten into any suitable sheet metal. Exterior surfaces are painted white and the interior ones, black, to minimize multiple reflection. The temperature sensor is mechanically ventilated by a small electrically driven fan, although the same design allows for direct mounting on a mast whose hollow arms at several heights provide an air path to a larger fan shared by several shielded thermometers. In this latter case, a vacuum cleaner may provide a useful high-volume air pump.

Particularly important features of this type of shield are that no horizontally incident radiation is able to reach the sensor directly and that with sufficiently high winds, adequate ventilation may be achieved without the use of the mechanical system because of the 360° horizontal opening. Fig. 4.7 includes a photograph of the shield shown diagrammatically in Fig. 1.6.

CHAPTER 2

Solar and terrestrial radiation

Experiment III has shown that at least two main spectral bands of radiation are received by bodies at the earth's surface. The short-wave, or visible radiation is received both directly from the sun and from scattering in the clouds and atmosphere as well as the earth's surface. Long-wave or infra-red radiation originates from polyatomic molecules, particularly H_2O and CO_2 in the atmosphere, and from solids and liquids such as constitute the earth's surface. Because of their high water vapour and liquid droplet concentration, clouds may also act as strong sources of infra-red radiation.

Radiation may be observed by noting the effect on a suitable physical property of a sensor, whose receiving surface absorbs a known fraction of a known bandwidth of the incident energy. Suitable properties of sensor materials that can be monitored are manifold, ranging from well-known basic quantities such as temperature to more complex photo-electric effects.

2.1. SPECIFIC INTENSITY AND RADIANT FLUX DENSITY

If the normal, $\hat{n}$, to the intercepting area and the direction of the specific intensity I are at an angle θ as shown in Fig. 2.1, then the flux density of radiation from a solid angle $d\Omega$ is given by:

$$dF = I \cos\theta \, d\Omega \qquad [2.1]$$

Regarding I and $\hat{n}$ as being in the plane of the paper it is clear that the flux through the paper would be zero. However, the flux density through a plane at right angles to the paper whose normal is also at right angles to $\hat{n}$, would be $I \sin\theta \, d\Omega$.

It follows that in general, the fluxes through unit areas with orthogonal normals in the x, y and z directions are related as follows:

$$dF_x^2 + dF_y^2 + dF_z^2 = (I d\Omega)^2 \qquad [2.2]$$

It is seen from equation [2.1] that the angular distribution of the intensity field has an important bearing on the integrated flux density. If the intensity is restricted to a single source of small angular extent, such as the sun, then the total flux is given by:

$$F = I \cos\theta \cdot \Delta\Omega \qquad [2.3]$$

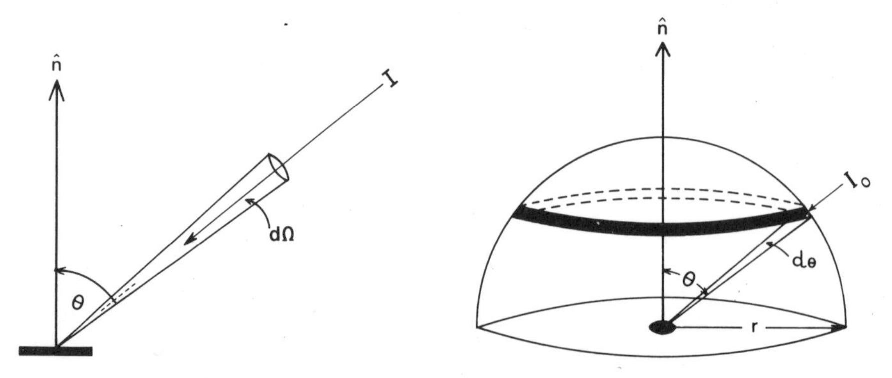

Fig. 2.1. The diagrammatic relationship between flux and intensity.

Fig. 2.2. Flux through a plane element of area from a hemisphere above.

where $\Delta\Omega$ is the finite, but small solid angle subtended by the source of radiation.

Although there is clearly no single general expression for the total flux density, an extremely important case is that of isotropic distribution of intensity over a hemisphere or 2π steradians. In this case as seen from Fig. 2.2:

$$F_o = 2\pi I_o \int_0^{\pi/2} \cos\theta \sin\theta \, d\theta$$

$$= \pi I_o \qquad [2.4]$$

where F_o is known as the isotropic flux density.

The incident visible radiation on completely overcast days and the radiation scattered from the earth's surface provide approximate examples of isotropic radiation.

Radiation detectors are ideally designed to respond linearly to the incident flux of a precisely known band of radiation. However, there are numerous problems which include variations in absorption of radiation with wavelength and angle of incidence.

2.2. RADIATION SCALES

The first three decades of this century saw the development of a large number of radiation instruments. Because of an over-indulgence of optimism, however, most of these instruments, including those subsequently employed as international standards, remained lacking in accuracy.

The two main traditional centres of radiation standards developed their own standard scales of radiation, namely, the Angström (in Europe) and the Smithsonian (in the U.S.A.). These scales, it must be emphasised, differed because of initially undetected instrumental errors, such as sensor edge effects. Today, it seems incredible that the two instruments, which acted as international standards during the same period of time, were never directly compared with each other. The scales which they represent are both incorrect by up to 2% and the magnitude of the correction required to bring older data to the present International Pyrheliometric Scale must depend somewhat on the intensities of incident radiation measured. It is assumed that the fractional correction required will not vary significantly for the range encountered in measuring the intensity of the direct solar beam.

On this basis, measurements made according to the original uncorrected Angström scale are increased by 1.5% and those according to the 1913 Smithsonian scale are decreased by 2% to bring all data as nearly as possible to a common level, that of the International Pyrheliometric Scale.

The great difficulties encountered in setting up an absolute radiation standard arise from:

(1) The uncertainty in the degree of blackness, i.e. absorption coefficient of the sensor surface.

(2) The difficulty in keeping heat loss from the sensor by unavoidable convective or conductive processes either negligible or accurately known.

2.3. THE FLUXES OF SOLAR AND TERRESTRIAL RADIATION

At the earth's surface, the sun at approximately 6000°K is seen almost as a "point source" subtending a solid angle of 0.68×10^{-4} steradians. The terrestrial flux on the other hand, originates from an entire hemisphere of 2π steradians and as a first approximation this latter radiation might be considered as isotropic. Considering, as a particular example, the earth to act as a radiator at 300°K, Stefan's Law indicates that the ratio of the solar to the terrestrial intensities would be in the ratio of the fourth power of the respective surface temperatures i.e. $(6000/300)^4 = 1.6 \times 10^5$. However, even with the sun at the zenith and unattenuated, the flux density ratios on a horizontal surface require inclusion of the ratio of the angular extents of the two sources, i.e. $(0.68/2\pi) \times 10^{-4}$ as a multiplier to the intensity ratio, yielding 1.73. The effect of attenuation of the solar beam in the earth's atmosphere as well as departure of the sun from the zenith is to reduce the value of this ratio. This implies that in general the solar and terrestrial fluxes of radiation, although in different wavelength ranges, are of comparable magnitude.

Planck's equation for the frequency distribution of black-body radiation can be used to show the solar and terrestrial bands of radiation are well separated, overlapping at a wavelength 4.7×10^{-6} m, where, for example, the equal flux contributions are only 0.25% that of the sun at 0.47×10^{-6} m, the wavelength of the solar maximum.

Experiment V. The measurement of radiation by a thermometric method

This is a more quantitative form of Experiment III and at once a development of Experiment II, in that three aluminium disks with respectively black, white and highly polished surfaces are exposed simultaneously. If the equilibrium temperatures are observed, it is theoretically possible to determine the incident fluxes of long- and short-wave radiation.

Fig. 1.1 is again referred to in discussing the balance of the several radiation fluxes in each case. An important assumption is that apart from radiation, the only other method of significant heat transfer available to the disks is by sensible heating of the air above their exposed surfaces. In this experiment, allowance must be made for the fact that all of the surfaces are imperfectly black or imperfectly white in either or both of the two spectral ranges under consideration. Thus for each disk a heat balance equation obtained from [1.10] and [1.11] applies as follows:

$$\omega(T - T_a) = \alpha S_i + \epsilon(L_i - \sigma T^4) \quad [2.5]$$

If all of these surfaces were ideal, then the following set of equations would be yielded after reference to Table 2.1a:

Black: $$\omega(T_B - T_a) = S_i + L_i - \sigma T_B^4 \quad [2.6]$$

White: $$\omega(T_W - T_a) = L_i - \sigma T_W^4 \quad [2.7]$$

Reflector: $$\omega(T_R - T_a) = 0 \quad [2.8]$$

where T_B, T_W and T_R are respectively the temperature of the black, white and reflector surfaces. In actual practice, the first set of observations on the temperature of the polished aluminium surface shows that equation [2.7] does not apply. Indeed further examination of the observed values of T_B and T_W indicates that values for α and ϵ similar to those shown in Table 2.1b would be more appropriate. These result in the following set of equations:

Black: $$\omega(T_{B'} - T_a) = 0.95 S_i + 0.95(L_i - \sigma T_{B'}^4) \quad [2.9]$$

White: $$\omega(T_{W'} - T_a) = 0.05 S_i + 0.95(L_i - \sigma T_{W'}^4) \quad [2.10]$$

Reflector: $$\omega(T_{R'} - T_a) = 0.05 S_i + 0.05(L_i - \sigma T_{R'}^4) \quad [2.11]$$

where $T_{B'}$, $T_{W'}$ and $T_{R'}$ represent the temperatures of imperfectly black, white and reflecting surfaces respectively.

Because the heat transfer coefficient ω is approximately independent of temperature, this term can be regarded as being equal for each of the surfaces, provided they are given an identical exposure to the wind. This means that if ω is determined by a set of observations based on one of the disks, say that with the polished aluminium surface, then S_i and L_i can be calculated from equations [2.9] and [2.10] with $T_{B'}$, $T_{W'}$ and T_a being the

TABLE 2.1

Radiative properties of perfect and imperfect surfaces

Surface	Short-wave absorptivity α	Long-wave emissivity ϵ
(a) *Ideal or perfect surfaces*		
Perfectly black	1.0	1.0
Perfectly white	0	1.0
Perfectly reflecting	0	0
(b) *Typical real surfaces*		
Black paint	0.95	0.95
White paint	0.05	0.95
Polished aluminium	0.05	0.05

only further observations required. For the determination of S_i alone, the air temperature does not need to be known, since:

$$S_i = \frac{\omega}{0.9}(T_{B'} - T_{W'}) \qquad [2.12]$$

The measurements are meaningful only if the incident radiation is constant. Further, in determining the reflecting disk's time constant and hence ω, it must be heated to a temperature exceeding $T_{R'}$ (which is best observed first) in the process of its cooling curve being plotted.

Finally, with ω, $T_{R'}$, T_a, S_i, L_i (and of course σ) all known, the validity of the suggested values of 0.05 for both α and ϵ in the case of the aluminium reflector can be checked. The apparatus is shown in Fig. 2.3, in which an additional white disk has been used to monitor the value of ω.

2.4. RADIATION INSTRUMENTS

Apart from pyrheliometers, which have only a small acceptance angle, a number of instruments have been developed which measure radiation from a solid angle of 2π or even 4π steradians. These instruments are designed to be sensitive in various wavelength regions. Usually they are made either to accept radiation from a given hemisphere or to register the net radiation by the cancelling effects of radiation incident on opposite sides of the sensor. Except for special purposes, the sensor consists of a flat horizontal plate and attempts to register the radiation incident on unit horizontal area at the location of the instrument. Basically, the sensor is similar to those employed in pyrheliometers. The incident radiation is either caused to set up a temperature gradient through the thickness of the sensor, or to result in differential heating across the horizontal surface as in the Moll thermopile used in the Linke and Feussner pyrheliometer and the Kipp pyranometer.

Fig. 2.3. Apparatus for the thermometric measurement of radiation. A group of three aluminium disks (partially machined out on their lower sides in order to reduce their mass and hence thermal time constants) have respectively black and white painted and polished metal surfaces. They are set into polyurethane foam and mercury-in-glass thermometers monitor their temperatures. On the left, a separate solid white-painted disk is used in an independent determination of the heat transfer coefficient ω.

Radiation instruments with relatively large acceptance angles require protection from the effects of convection. Either of two methods is usually adopted. One approach is to make natural convection negligible by means of a blown draft of air across the sensor surface. In the case of net radiometers, the air stream has to be adjusted to equalize convective heat transfer over both the upper and lower surfaces. If the draught is kept constant, this type of instrument can be calibrated. The alternative solution is to provide the sensor surface with a suitable dome which does not absorb any radiation in the band being measured.

Short-wave or visible light radiometers are known as pyranometers and accomplish doming without difficulty by using glass. This material absorbs long-wave radiation, so that unless the dome and the sensor body are at the same temperature, re-radiation from the dome results in a net effect on the instrument reading. Good thermal contact between the dome and the short-wave radiometer body is thus essential. Performance is improved by the use of multiple domes, two only usually being chosen, as otherwise multiple reflection becomes serious. Long-wave or all-wave radiometers cannot use glass, of course, and can only be domed by very thin polythene. This type of dome must be kept inflated by a source of dry compressed air or nitrogen, a principle which is adopted in the radiometer designed by P. Funk of the Australian C.S.I.R.O. and since improved by W. Beerli of Swissteco. The Gier and Dunkle all-wave radiometer, of earlier design, has no dome and employs a collimated draught of air. The chief disadvantage of the latter type of instrument is that even in moderately strong winds allowance must be made for calibration changes.

Short-wave radiometers having black sensor surfaces, as have those made by Kipp, Swissteco and now Eppley, essentially differ from the all-wave instruments only in the type of protection from convection. The old Eppley pyranometer differs here in having a black disk surrounded by white and black annular areas. Because the white surfaces act as a black body in the long-wave region, a temperature difference between the black and white portions of the sensor is caused only by short-wave radiation. Apart from short- and all-wave radiometers, a third type occasionally encountered has a white sensor surface, which reflects most of the incident short-wave radiation, and an inflated polythene dome, thus being essentially a long-wave detector. Unfortunately, since the sun is such an intense source of short-wave radiation, the imperfect rejection of visible radiation by this type of long-wave radiometer means that unless the sun is occulted, a considerable error is caused. The performance of the instrument even on overcast days is probably unreliable. It is thus more usual to measure all- and short-wave radiation on separate instruments, whereby long-wave energy may be computed if required.

Imperfections in the degree of blackness (or whiteness) of sensor surfaces can be compensated for by means of special techniques. For example, in all-wave radiometers it may be found that the response is greater for a given flux in the short-wave spectral region than for the same flux in the long-wave. Equality of responses in this case can be achieved by painting white a suitable fraction of the otherwise black sensor surface, thus partially reducing the short-wave sensitivity only by a controlled amount.

It is possible to construct a satisfactory long-wave sensor by compensating for the undesired partial short-wave response caused by the imperfectly white surface (which can absorb from 5% to 10% of the incident short-wave radiation) by means of connecting a smaller separate thermopile with black sensor surfaces of suitable area in opposition to the white-covered thermopile. This has been done by J. Mitchell of Solar Radiation Instruments.

Globally sensitive radiometers with plane sensors may exist in two forms. One is suitable for measuring incident radiation from the single hemisphere seen by the sensor surface; the other is designed to measure the net exchange of radiant energy through the plane of the instrument. When sensitive to short-wave radiation only, the net radiometer is known as a short-wave balance meter.

All radiometers with plane sensors, when facing radiation from a source of very small angular extent, which will be referred to as "point source" or direct beam radiation, have a calibration "constant" depending on the angle of incidence. This is particularly serious for radiation at grazing incidence, and is due to the failure of Lambert's Law to apply to the imperfectly black surface of the sensor. Thus radiometers used in polar regions must be particularly carefully calibrated. One important result of this angular dependence is that the calibration varies significantly for point source and diffuse radiation. Thus it is almost impossible to draw conclusions of high accuracy from readings obtained on partially overcast days. It is clear that the necessary corrections can be more easily applied to instruments sensitive to a single hemisphere rather than net instruments, which monitor radiation over both upper and lower hemispheres and hence may be exposed to direct radiation in one hemisphere and to isotropic in the other. Hence short-wave balance or net short-wave radiation is more accurately measured with two separate pyranometers, each used with the most suitable calibration "constant", In the case of the all-wave net radiometer, the convenience of having an instrument capable of giving a direct reading of the net total radiant energy exchange is usually regarded as outweighing the calibrational disadvantages. The all-wave net radiometer can be calibrated particularly accurately for overcast days and night-time use, both conditions involving diffuse radiation being incident on both sides of the sensor. When the lower sensing surface of an all-wave net radiometer is covered by a cup with blackened interior whose temperature can be measured, the flux of radiation incident on the domed face can be calculated. The reading in this condition is made up of the thermopile output and the flux from the blackened cup, so that a zero thermopile e.m.f. does not imply no incident radiation.

The complete observation of the main fluxes of radiation at the earth's surface may be accomplished in general by five separate instruments:

(1) Pyranometer, to measure globally incident short-wave radiation, S_i.

(2) Pyrheliometer, to measure only direct short-wave solar radiation, Z_i, and the angle of the sun from the zenith, θ. Alternatively, a second pyranometer equipped with a disk to occult the direct solar beam can be used to provide a reading of $S_i - Z_i \cos\theta$.

(3) Pyranometer, to measure scattered and reflected (or outgoing) short-wave radiation, S_o.

(4) Net radiometer, to measure net radiant energy of all wavelengths, A_N.

(5) Net radiometer with reference cup, to measure incident all-wave radiation, A_i.

From the five quantities Z_i, S_i, S_o, A_i and A_N, it is possible to obtain:

(6) The incident diffuse short-wave sky radiation, $C_i = S_i - Z_i \cos\theta$.

(7) The short-wave albedo of the surface below, $\beta = S_o / S_i$.

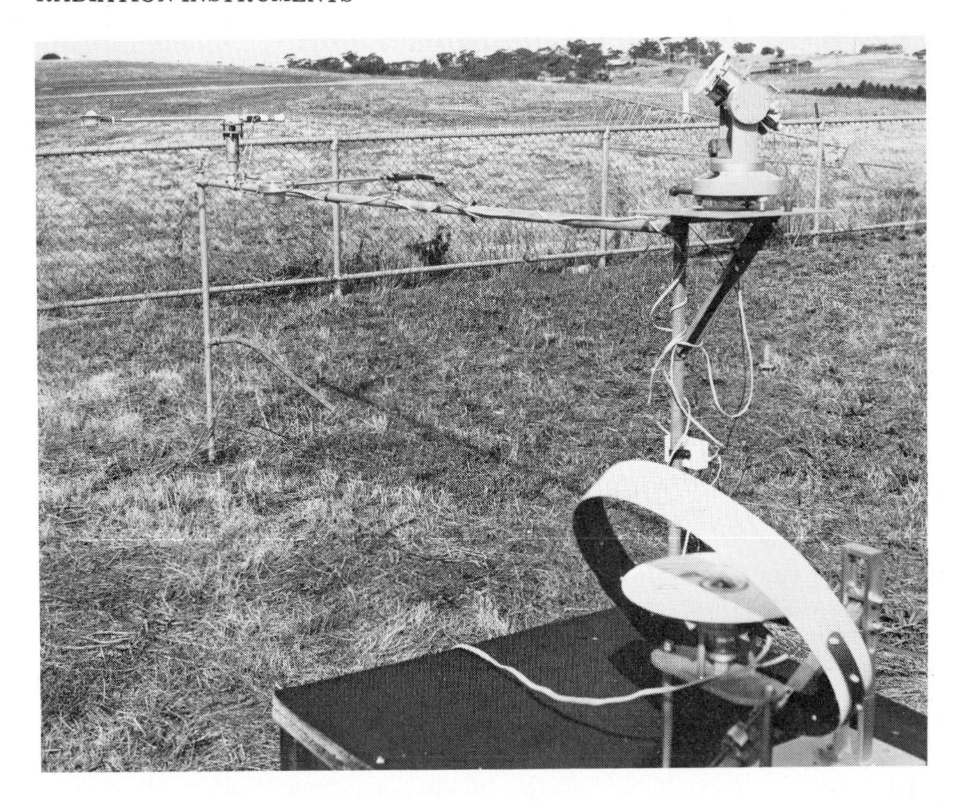

Fig. 2.4. Radiation instruments. From left to right on the tabular stand are seen respectively an albedometer (consisting of two pyranometers mounted back to back so that one registers incoming and the other outgoing radiation), a net radiometer (with the lower polythene hemisphere replaced by a metallic interiorly blackened metal cup equipped with a thermocouple) and a Linke-Feussner pyrheliometer.

(8) The outgoing all-wave radiation, $A_o = A_i - A_N$.
(9) The incident long-wave radiation $L_i = A_i - S_i$.
(10) The outgoing long-wave radiation $L_o = A_o - S_o$.

The photograph in Fig. 2.4 shows various radiation instruments mounted on a simple tubular frame in a manner suitable for continuous observation.

Experiment VI. Calibration of a pyranometer against a pyrheliometer

Experience has shown that the simpler principles and greater robustness involved in the construction of a pyrheliometer lead to its calibration holding more reliably than domed instruments, as well as those whose sensitive surfaces are exposed more continually to deteriorating influences. Accordingly, a suitable pyrheliometer is often used as a secondary standard and brought into use mainly for calibration checks of other instruments such as pyranometers and all-wave net radiometers.

Calibration of a pyranometer illustrates the basic procedure involved which, for example, may also be used in providing all-wave net radiometers with a short-wave calibration. A clear day is chosen and the solarimeter is exposed to obtain a reading or electrical response, ϵ_s (mV), for the total incident or global short-wave radiation, S_i. If the unknown calibration constant of the solarimeter is k_s (mV mW^{-1} cm^{-2}), then if the pyranometer sensor is assumed to be perfectly black:

$$\begin{aligned} \epsilon_s &= k_s S_i \\ &= k_s(Z_i \cos\theta + C_i) \end{aligned} \quad [2.13]$$

After this, the sun is shaded from the instrument by using an occulting disk subtending the least possible solid angle, enabling a reading, ϵ_c, for the sky radiation alone to be obtained as follows:

$$\epsilon_c = k_s C_i \quad [2.14]$$

The response that the radiometer would show to the sun alone is thus:

$$\begin{aligned} \epsilon_z &= \epsilon_s - \epsilon_c \\ &= k_s Z_i \cos\theta \end{aligned} \quad [2.15]$$

where it has been assumed that the same calibration constant can be used for the direct and the diffuse beam. Although this is usually incorrect, the error introduced is negligible because of the small magnitude of the diffuse component compared to the direct beam. The pyrheliometer is then faced into the sun to give both the solar zenith angle θ and through its known calibration constant k_z and observed response ϵ_o enable a determination of Z_i:

$$\epsilon_o = k_z Z_i \quad [2.16]$$

From equations [2.15] and [2.16] the pyranometer calibration then follows as:

$$k_s = \frac{\epsilon_s - \epsilon_c}{\epsilon_o \sin\theta} k_z \quad [2.17]$$

A number of these sets of observations should of course be carried out in order to obviate errors introduced by otherwise undetected short-term changes in atmospheric transparency.

2.5. LAMBERT'S OR THE COSINE LAW

The flux of radiation from a surface element of a perfectly black radiator is proportional to the cosine of the angle between the direction of emission and the normal to the surface. In converse form, a similar statement holds for absorption as well. Since radiometer sensor surfaces are ideally perfectly black absorbers it is important for every radiometer to be examined for departure from this behaviour.

Several factors in the design of solarimeters, not only the imperfect cosine response of the sensor surfaces, can combine to produce unexpectedly large variations in the calibration factor which careful workers always check as a function of both solar elevation and azimuth relative to the instrument.

The Moll thermopile, for example, while perfectly well suited to normal incidence measurement such as in the Linke-Feussner pyrheliometer, is not completely satisfactory when used in pyranometers or short-wave radiometers because the upper surface in consisting of several adjacent ribbon thermojunctions, is relatively uneven. This unevenness becomes increasingly serious as grazing angles of incidence are approached.

Radiometers having square rather than circular thermopile surfaces can also suffer from azimuthal variations in sensitivity. This can be especially serious in pyranometers at low angles of solar incidence when the inner glass dome surfaces can focus a concentrated pattern of light on to the edge of the thermopile to an extent which varies with the horizontal orientation of the square thermopile surface.

Experiment VII. Cosine response of a radiometer

Although it is possible to observe the response caused by an artificial light source, special equipment is needed which keeps the light source intensity constant while it is moved around the instrument under test, since the latter must be kept horizontal. It is preferable to use the sun as a "point source", and to perform repeated calibrations of the instrument during the course of several hours on a clear day. The greatest change occurs when the sun draws near to the horizon. For this reason, this more exacting type of calibration is most important for radiometers being used in polar regions. The same techniques are applicable to all net and hemispherically sensitive (or global) instruments.

The experiment is ideally performed by a group which can arrange for observations to be continued over the greater part of the day. If a clockwork-driven equatorial mount is available for the pyrheliometer, it is a simple matter to obtain calibrations for the full range of angles of incidence from the horizontal to the maximum altitude as determined by the elevation of the sun at solar noon. The latter, of course, depends both on the latitude of the site and time of year.

2.6. DIRECT BEAM AND DIFFUSE CALIBRATIONS FOR A RADIOMETER

Let the response of a radiometer to a flux F of direct beam (or "point source") radiation be $\epsilon_z(\theta)$ where θ is the zenith angle of the source, for example the sun, and let $k_z(\theta)$ be the calibration factor observed at that angle. Thus if the solar flux at normal incidence is Z_i, then:

$$F = Z_i \cos\theta \quad [2.18]$$

and:

$$\epsilon_z(\theta) = k_z(\theta) Z_i \cos\theta \quad [2.19]$$

Although the calibration factor will be found to have some temperature dependence, most importantly it will be proportional to the absorptivity of the sensor element for a source at an angle θ to the normal.

Assuming now that the same total flux F is incident under perfectly diffuse conditions as might be approximated on an overcast day, the intensity I_o is isotropically distributed over a hemisphere (as shown in Fig. 2.2) and $F = \pi I_o$. Thus for a given infinitesimal band of sky, concentric about the zenith at an angle θ, the flux is made up of contributions such that:

$$F = 2\pi I_o \int_0^{\pi/2} \sin\theta \cos\theta \, d\theta$$

The response of the same radiometer follows as:

$$\epsilon_c = \int_0^{\pi/2} d\epsilon_c(\theta) = 2\pi I_o \int_0^{\pi/2} k_z(\theta) \sin\theta \cos\theta \, d\theta \quad [2.20]$$

If $k_z(\theta) = k_z(0)$, a constant, then the result is trivial:

$$\epsilon_c = \pi I_o k_z(0) = k_z(0) F \quad [2.21]$$

as for the direct beam.

However, departures from the cosine law are to be expected for radiometer sensor surfaces, so that more generally the isotropic response is given by:

$$\epsilon_c = 2F \int_0^{\pi/2} k_z(\theta) \sin\theta \cos\theta \, d\theta$$

$$\therefore \frac{\epsilon_c}{\epsilon_z(\theta)} = \frac{2 \int_0^{\pi/2} k_z(\theta) \sin\theta \cos\theta \, d\theta}{k_z(\theta)} = \frac{k_c}{k_z(\theta)} \quad [2.22]$$

where k_c is the calibration factor for isotropic radiation.

Many manufacturers of radiation instruments quote only a single calibration factor that has been determined for normal incidence of the direct solar beam, for which:

$$\frac{\epsilon_c}{\epsilon_z(0)} = \frac{2\int_0^{\pi/2} k_z(\theta)\sin\theta\cos\theta\,d\theta}{k_z(0)} = \frac{k_c}{k_z(0)} \qquad [2.23]$$

Preparatory to accurate radiation measurements, it is most important to measure the ratio in equation [2.23] since 5% variations are not uncommon. An extension of Experiment VII is thus to evaluate equation [2.23] using the measured values of $k_z(\theta)$. This is best done graphically or numerically.

Experiment VIII. Dependence of the albedo on solar elevation

A number of terrestrial surfaces have a structure that might be expected to contribute to the phenomenon of diurnally varying albedo. Water surfaces are particularly complex in that the orientation of waves of various frequencies and amplitudes result in temporal and spatial variations in reflectivity (and polarization). Areas of straight grass stalks are more easily analysed and it is clear that variations in the albedo occur between occasions when the sun near the zenith, directly illuminates dark soil spaces between plants and later, nearer to the horizon, is intercepted and scattered only by stalks and leaves.

Such observations require several hours in order to secure adequate variation of solar angle but if two pyranometers have been carefully calibrated they can be connected to a suitable recorder without requiring the continued attendance of an observer.

Because most of the earth's surface is structured by plants (including trees) and waves on the oceans, the consequences of these albedo variations is of great importance.

During the measurement of albedo on a clear day, it is appropriate to use the "point source" or direct beam calibration for the pyranometer monitoring incoming flux so that:

$$\begin{aligned}\epsilon_z(\theta) &= k_z(\theta)(Z_i\cos\theta + C_i)\\ &= k_z(\theta)S_i\end{aligned}$$

This approximation is acceptable because the diffuse component of sky radiation $C_i \ll Z_i \cos\theta$ for $\theta < 80°$.

On the other hand, the upward scattered flux from the earth's surface might be regarded as diffuse, making the use of the diffuse calibration more appropriate:

$$\epsilon_c = k_c S_o$$

The albedo is then given by:

$$\beta = \frac{S_o}{S_i} = \frac{\epsilon_c k_z(\theta)}{\epsilon_z(\theta) k_c} \qquad [2.24]$$

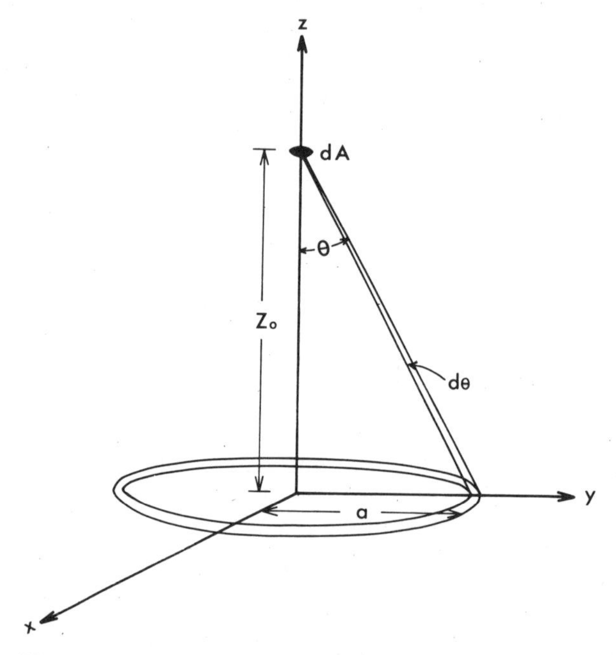

Fig. 2.5. Construction for calculating the radiated flux through a plane element of area dA, mounted parallel and symmetrically above a finite, plane, circular radiating surface of radius a.

2.7. RADIATION MEASUREMENTS OVER FINITE PLANE SURFACES

The accuracy to be expected from data yielded from a site chosen for albedo measurements depends on the extent and uniformity of the area. If the anisotropies of the surface are three dimensional such as are formed by plants, terrestrial undulations and oceanic waves, then the height of the downward-facing instrument also influences the magnitude of the reading for the albedo.

However, a simple two-dimensional horizontal surface will be considered here.

If the area dA in Fig. 2.5 represents the sensor of an inverted pyranometer mounted at a height z_0 over an infinitely extending horizontal surface, then the flux density, F, of radiation at the instrument is made up of a series of contributions originating from concentric annuli as shown:

$$F = 2\pi \int_0^{\pi/2} I_\theta \sin\theta \cos\theta \, d\theta \qquad [2.25]$$

The simplest case is that of an isotropic distribution of intensity I_o from an infinitely extending plane radiator (or scatterer), so that $F = \pi I_o = F_o$, simply indicating that the sensor intercepts the isotropic flux.

Usually, however, the area available for albedo observations is restricted. As an example, if the area is circular and of radius a, then the radiant flux density intercepted by a detector mounted at a height z_o is:

$$F_a = 2\pi I_o \int_0^{\sin^{-1}(a/\sqrt{a^2+z_0^2})} \sin\theta \cos\theta \, d\theta$$

$$= \frac{a^2}{a^2 + z_0^2} F_o \qquad [2.26]$$

Furthermore, it can readily be shown that if the area outside the circle of interest is associated with a different isotropic intensity of nI_o, then it contributes a flux density at the instrument of:

$$n(F_o - F_a) = n\pi I_o \frac{z_0^2}{a^2 + z_0^2}$$

The total flux density received by the sensor is now:

$$F_a' = F_a + n(F_o - F_a)$$

$$= \left(\frac{a^2 + nz_0^2}{a^2 + z_0^2}\right) F_o \qquad [2.27]$$

Two further examples illustrate the importance of evaluating the above parameters prior to accepting uncorrected results from a given site.

For $n = 0$, i.e. a non-radiating area outside the circle of interest:

$$F_a' = \frac{a^2}{a^2 + z_0^2} F_o$$

and for $n = 2$, i.e. the exterior area having twice the albedo of the inner circle:

$$F_a' = \frac{a^2 + 2z_0^2}{a^2 + z_0^2} F_o$$

It follows that if prescribed that independently of other instrumental errors, the actual site should not generate further inaccuracies in excess of 2%, then in both of the above cases the requirement of $|F_a - F_a'| \leqslant 0.02$ would imply a circular area having a minimum radius of approximately 7 m, assuming that the sensor is mounted at a height of 1 m.

Experiment IX. Measurement of the albedo over finite surfaces

The albedo (and also the net radiation) is a quantity that is frequently of necessity measured over a limited area. For example, agrometeorological investigations may call for the assessment of plant community albedos from data obtained from relatively small test plots. The effect of certain engineered surfaces such as tarmac or cut grass on the radiation balance may also constitute a problem for investigation over limited surface areas.

Equation [2.27] can be used to calculate the albedo of an infinite surface for which only finite sampling is available. In a simple demonstration of the method, the albedo of a large horizontal surface such as presented by a field of grass is measured. It is assumed that the area is sufficiently large to be regarded as infinite in extent.

A circular area of contrast should then be formed by spreading flour or lime for example and the effective albedo observed from a point of known height directly above the centre of the cirle, whose radius should also be measured carefully. Measurements should be made with the pyranometer at two or three different heights.

Experiment X. The measurement of long- and short-wave radiation fluxes

A pyranometer may be used to observe total incident short-wave energy as well as that reflected and scattered from the earth's surface. An all-wave net radiometer can be used either in its basic form or adapted to measure all-wave radiation from one hemisphere only. In the latter form, a black body in the form of a cup, reflecting on its outer surface, is mounted over one sensing surface of the instrument. Net radiation is thus measured over a surface whose radiation flux density is known through measurement of its temperature by means of a thermocouple, for example. There is thus no fundamental difference between the two forms of all-wave radiometer between which most manufacturers allow for a simple conversion. Were it possible to measure the radiation from the earth's surface below an all-wave net radiometer by a simple measurement of temperature alone, the same result would be achieved as in using the radiometer in its single hemispherical adaptation. The actual reading of the instrument in the latter case must be combined with the calculated radiation from the reference surface below.

The all-wave radiometers should be used to establish net, A_N, and incident, A_i, radiation. The pyranometer may be used to establish incident S_i, and reflected, S_o, short-wave radiation, from the ratios of which the short-wave albedo β follows.

It is interesting to establish outward long-wave radiation, L_o, by two methods. Firstly, from the instrumented readings, since:

$$L_o = A_o - S_o \qquad [2.28]$$

where the values of A_o and S_o are obtained by facing both of the relevant instruments toward the ground. Alternatively, the value of A_o may be calculated from net and incoming all-wave observations: $A_o = (A_i - A_N)$.

A second independent method involves a direct temperature measurement of the surface below the instruments. When the surface is relatively well defined as in the case of bare soil, a suitable thermometer can be embedded just below the soil surface but because of the large temperature gradients which frequently prevail at this boundary, great care needs to be taken.

If the experiment is repeated over short thick grass, for example, there is great difficulty in deciding on which level of the permeable surface presented has the most typical radiating temperature.

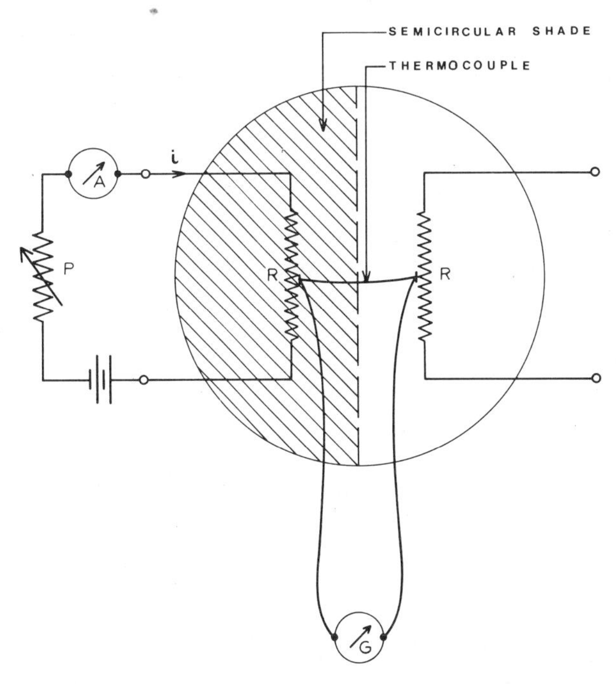

Fig. 2.6. Cross-sectional diagram of the Angström-type compensation pyrheliometer. The temperature difference between identical metal strips of resistance R is monitored by glued-on thermocouples connected to a galvanometer G. The electrical current i through the shaded strip is measured by the ammeter A and controlled by potentiometer P.

Experiment XI. A basic pyrheliometer

Angström's electrical compensation pyrheliometer, although usually employed as a secondary comparison instrument, can also be used absolutely.

The detector assembly consists of two thin blackened strips, one of which is shaded and the other exposed to the direct solar beam. Thermocouple junctions adhering to (but electrically isolated from) the centre of the two strips allow the temperature difference to be observed and brought to zero by the regulated passage of an electrical heating current through the shaded strip as shown in Fig. 2.6. In this latter condition, the electrical power dissipated in the shaded element may be regarded as equal to the solar flux absorbed by the exposed strip.

If the electrical resistance of the physically identical strips is R and their area is A, then the power dissipated per unit area is Ri^2/A. Thus if i is the current required to equalize the strip temperatures and α is the radiant absorptivity of the exposed strip, then, on normal incidence, the solar flux density is given by:

$$Z_\mathrm{i} = Ri^2/\alpha A \qquad [2.29]$$

In principle this type of instrument can be constructed readily. For instructional purposes some accuracy may be sacrificed and any suitable electrical resistive strips and thermocouple materials can be used. The two metal strips can be mounted on conducting rods set into an insulating disk and collimation of the solar beam can be ensured by means of a blackened metal tube which can be held by a simple clamp in the orientation required.

The function of the basic pyrheliometer can be checked either by comparison with a suitable standardizing instrument (such as a Linke-Feussner pyrheliometer) or by an actual observation of the solar constant as described in the following experiment. In the latter technique, the extra-terrestrial solar flux density, just outside the attenuating influence of the atmosphere, is regarded as known and effectively used as a standardizing source.

The zenith angle of the sun can be measured with sufficient accuracy using a simple assembly of a spirit level, protractor and pinhole sighting scope.

2.8. THE EXTRA-TERRESTRIAL SOLAR FLUX

The flux density of solar radiation normally incident on a flat surface at the mean distance of the earth from the sun with the atmosphere being considered absent, is often known as the "solar constant". For many practical purposes, this is regarded as the flux density of solar radiation at the outer limit of the earth's atmosphere.

Variations in this latter flux arise from the varying distance of the earth from the sun as well as fluctuations which may be up to $\pm 1.5\%$ in solar emissive power.

The sun being closest to the earth in January, results in a significant lack of symmetry between the total annual irradiation of the northern and southern hemispheres. Corrected for the mean solar distance, the currently accepted value is $136\,\mathrm{mW\,cm^{-2}}$.

Experiment XII. Determination of the solar constant

Because of dispersion and differential absorption, an accurate determination of the solar constant would involve a complete scan of the incident spectrum. A combination of both varying absorption, refraction and scattering effects with local atmospheric conditions as well as altitude and angle of incidence make a precise analysis difficult.

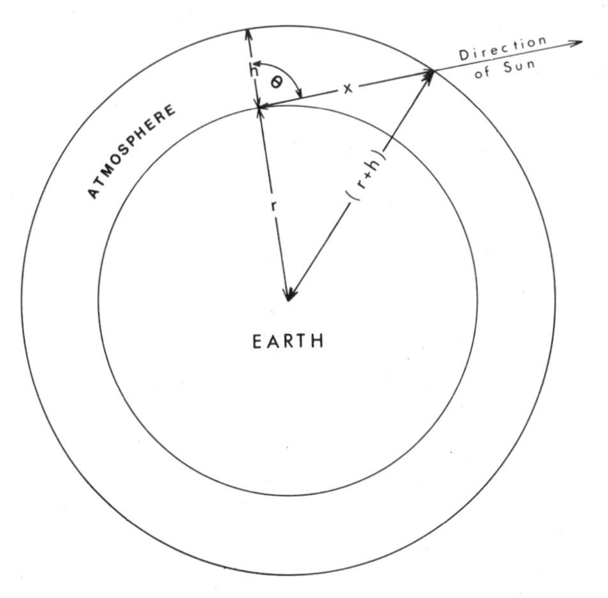

Fig. 2.7. A simplified geometry of the sun, earth and atmosphere.

On the basis of simplifying assumptions, the diagram shows the type of geometry involved. The effective extent of the atmosphere is given by h, so that at an angle θ from the zenith, the thickness of atmosphere traversed by solar radiation is x. From Fig. 2.7 it can easily be shown that:

$$x = -r\cos\theta + \sqrt{r^2\cos^2\theta + h^2 + 2hr} \qquad [2.30]$$

where r is the radius of the earth.

For the trivial case of $\theta = 0$, equation [2.30] gives $x = h$. For $\theta = 90°$, $x = \sqrt{h^2 + 2hr}$. The practical application of this equation thus depends on postulating an effective atmospheric thickness in the same units as the earth's radius. For small values of θ, $x \approx h\sec\theta$.

It will have become increasingly clear that the use of a standard set of tables connecting x and θ is desirable. Table 2.2 shows the effective mass of air, M, relative to $\theta = 0$ as a function of angle. In the sense of equation [2.30], $M = x/h$. With a surface atmospheric pressure of p mbar the value of M should be modified to $M_p = M \cdot p/1015$.

The experiment should be performed by a group taking turns to observe the solar radiation intensity with a suitable pyrheliometer at regular time intervals overlapping noon. The results should be plotted graphically as a

TABLE 2.2

Relative mass of air, M, penetrated by radiation with the sun at an observed angle, θ, from the zenith compared to secθ

θ	Secθ	M	θ	Secθ	M
0	1.000	1.000	66	2.459	2.447
5	1.004	1.004	68	2.669	2.654
10	1.015	1.015	70	2.924	2.904
15	1.035	1.035	72	3.236	3.209
20	1.064	1.064	74	3.628	3.588
25	1.103	1.103	75	3.864	3.816
30	1.154	1.154	76	4.134	4.075
35	1.221	1.220	77	4.445	4.372
40	1.305	1.304	78	4.810	4.716
42	1.346	1.344	79	5.241	5.120
44	1.390	1.389	80	5.76	5.600
46	1.440	1.438	81	6.39	6.18
48	1.494	1.492	82	7.19	6.88
50	1.556	1.553	83	8.21	7.77
52	1.624	1.621	84	9.57	8.90
54	1.701	1.698	85	11.47	10.40
56	1.788	1.784	86	14.34	12.44
58	1.887	1.882	87	19.12	15.36
60	2.000	1.995	88	28.65	19.80
62	2.130	2.123	89	57.30	27.00
64	2.281	2.274	90	∞	39.70

function of effective air mass M and extrapolated to zero to obtain the solar constant.

CHAPTER 3

Air and water vapour pressure

The atmosphere has several constituent gases including mono- and poly-atomic molecules. Most of these are present in some constant proportion, for example nitrogen and oxygen, because there is neither a significant source of production or sink, nor liquid or solid phases of the substance present in the terrestrial system. Although plants and animals are involved in oxygen cycles the quantities in use are negligible when compared to those present in the atmosphere. Water vapour on the other hand, because it co-exists with both liquid water and ice in many widespread forms over the earth and in the atmosphere undergoes enormous variations in concentration in the latter medium.

Carbon dioxide, being a by-product of transportation as well as of combustion, including both fires of natural materials such as forests and many industrial processes and a requirement for plant photosynthesis undergoes marked variations in atmospheric concentration in both space and time. In each hemisphere seasonal oscillations can be observed which are related to plant functions. However, with a partial pressure of about 0.3 mbar, carbon dioxide is a less massive constituent of the atmosphere than water vapour, whose saturation vapour pressure at 25° C is approximately 100 times as great.

Except perhaps occasionally in polar regions, carbon dioxide does not change phase on earth. There is no reservoir of liquid or solid material available as in the case of water whose pressure in the atmosphere, even under saturation conditions, is strongly temperature dependent. Excess carbon dioxide dissolves in the sea resulting in a natural, only slowly changing, overall equilibrium.

3.1. ATMOSPHERIC PRESSURE

The measurement of atmospheric pressure is of considerable importance in meteorology being the basis of the daily meteorological map. Pressure is defined as the force per unit area exerted by a medium, and as it is isotropically directed, no specification of orientation is required. If the acceleration of gravity, g, is the only acceleration present and if m is the total mass of the atmosphere overlying area A, the pressure, p, is given by:

$$p = mg/A \qquad [3.1]$$

In the c.g.s. system of units, the pressure is expressed in terms of dyne cm^{-2}. Since normal atmospheric pressure is approximately 10^6 dyne cm^{-2} this latter quantity is defined as 1 bar. The usual meteorological unit is the millibar equal to 10^3 dyne cm^{-2}.

The chief interest in meteorology lies in pressure gradients, both in terms of differences in time and between one place and another. There is little interest in small differences which may be due to differences in the acceleration of gravity at various locations. This varies with latitude and elevation and is also affected by the mass of mountains. By international agreement, observations are reduced to standard gravity, the value 980.62 cm sec^{-2} prevailing at sea level at 45° latitude being specified. At most stations the gravity correction is small enough to allow it to be considered a constant value added to all barometer readings.

In order to be able to compare readings at stations having different elevations, it is necessary to reduce all observations to a common level, usually that of the sea, although this practice is questionable over great continental plateaux. Whenever the temperature distribution with height has unforeseen features, such as occur when valleys are filled with layers of air cooler than the surrounding mountains, the pressure reduction necessary for meteorological maps becomes an extremely inaccurate process. This is unfortunate as many instruments have been devised for the accurate measurement of local pressure.

Atmospheric pressure is measured by means of a wide range of instruments called barometers. The various forms usually rely on a fluid or solid transducer responding to a force against an airtight material separating atmospheric pressure from some reference pressure. In the case of the mercury barometer, this reference pressure is a near vacuum. Exceptions involve the monitoring of some pressure-sensitive physical property such as the boiling point of a suitable liquid.

3.2. LIQUID COLUMN BAROMETERS

The balancing of the atmosphere against a column of mercury has become so common a procedure that the length of the supported mercury column has become an accepted unit of pressure.

Basically this device was discovered by Torricelli in 1643. The cross-sectional area of the column does not affect the reading unless the tube is so narrow that capillary forces become important. The mercury also undergoes thermal expansion, i.e. changes in density. Since the weight of mercury over unit area is the significant quantity, a standard temperature of 0° C is chosen for the reduction of all readings.

If ρ_T is the density of mercury at a temperature T, then the pressure exerted by a mercury column of height h, uniform cross-sectional area A and mass m is:

$$p = mg/A = \rho_T gh \qquad [3.2]$$

Taking $g = 980.6\ \text{cm sec}^{-2}$ and $\rho_0 = 13.60\ \text{g cm}^{-3}$ it is seen that 760 mm of mercury at 0°C are equivalent to 1013.5 mbar.

Two types of mercury barometer are in general use — the Fortin type and the fixed cistern or Kew type. In the Fortin barometer the level of the mercury reservoir is adjusted to the zero of the column scale prior to each reading. This is achieved by a kid skin bottom making the volume of the cistern adjustable so that varying volumes of mercury in the column can be compensated for. In the Kew type of barometer no adjustment is necessary as the fall of mercury in the cistern accompanying a rise in the column is taken into account by a compensated scale.

All mercury barometers are intended for accurate measurements, and thus incorporate sliding verniers to give at least tenths of a millimetre. The scale is intended for reading at the highest point of the mercury meniscus. Temperature corrections if not available with a particular instrument can be found in the *Smithsonian Meteorological Tables* (6th ed., p. 165).

Experiment XIII. A short water barometer

The sensitivity of a liquid barometer is inversely proportional to the density of liquid in the column. This implies that a water barometer would be useful in the study of short-term pressure fluctuations whose amplitude would escape detection by conventional mercury barometers. However, since the pressure equivalent of 760 mm of mercury is 10.34 m of water, it is clear that a relatively cumbersome instrument would result. Nevertheless by using thick-walled plastic tubing and a three-storey building for example, atmospheric pressure can be determined by measuring the vertical extent of the water column. For reasonable precision, the variation of water density with temperature must be allowed for, as well as water vapour pressure.

A more compact water barometer uses a shorter length of tubing connected to an airtight metal container at the upper end and to a suitable reservoir of water at the lower extremity as shown in Fig. 3.1. By carefully heating the metal container a small volume of the air can be expelled by bubbling out through the reservoir so that during the process of returning to room temperature, a column of liquid will be sucked into the tube. The height of the column of water formed gives an indication of the difference between atmospheric pressure p_A and that of the air p_a plus the saturated water vapour e_s within the container, so that:

$$p_A = \rho gh + p_a + e_s$$

$$\therefore p_A = \rho gh + \frac{nR}{V}T + e_s \qquad [3.3]$$

which follows from applying the ideal gas equation to the air volume:

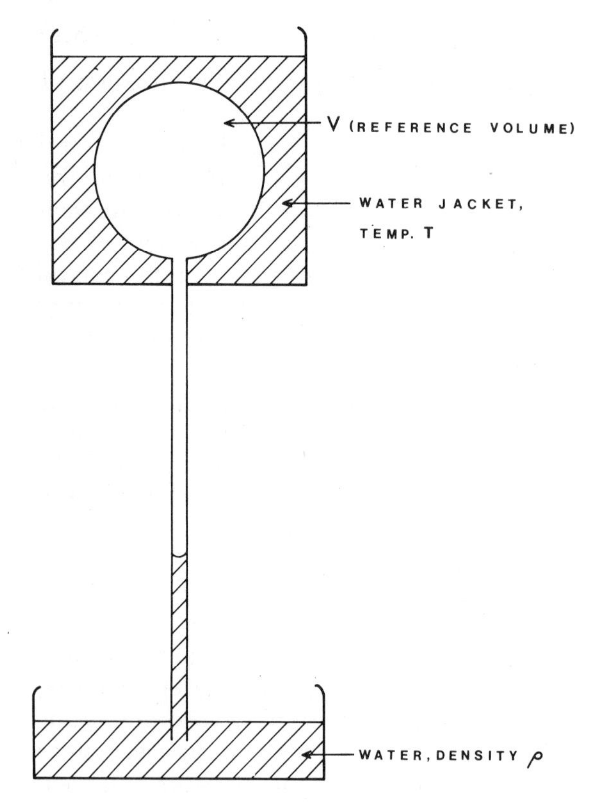

Fig. 3.1. Diagram of a short water barometer.

$$p_A V = nRT \qquad [3.4]$$

where ρ is the density of water in the column of length h, g is the acceleration caused by gravity and n, V and T are the number of moles, volume and absolute temperature of the air in the reference pressure container and R is the universal gas constant. Because the total water vapour content varies, it is not included in the ideal gas expression which is used for dry air alone.

It is evident from equation [3.3] that the column height is not only dependent on atmospheric pressure but also on the temperature and water vapour pressure. This set of interdependences can be explored by differentiating equation [3.3] for constant atmospheric pressure:

$$0 = \rho g - \frac{nR}{V}\left(\frac{T}{V}\frac{\mathrm{d}V}{\mathrm{d}h} - \frac{\mathrm{d}T}{\mathrm{d}h}\right) + \frac{\mathrm{d}e_s}{\mathrm{d}h}$$

$$\therefore \frac{nR}{V} = \left(\rho g + \frac{\mathrm{d}e_s}{\mathrm{d}h}\right) \Big/ \left(\frac{T}{V}\frac{\mathrm{d}V}{\mathrm{d}h} - \frac{\mathrm{d}T}{\mathrm{d}h}\right) \qquad [3.5]$$

TABLE 3.1

Saturation vapour pressure over ice and water in mbar as a function of temperature. For temperatures below freezing, the upper and lower figures are for ice and supercooled water respectively (note: 1 mm Hg = 1.3336 mbar)

Temperature (°C)	Add differences to temperature									
	0	1	2	3	4	5	6	7	8	9
−30	0.38	0.42	0.47	0.52	0.57	0.63	0.70	0.77	0.85	0.94
	—	—	—	—	—	—	—	—	—	—
−20	1.03	1.14	1.25	1.37	1.51	1.65	1.82	1.99	2.18	2.38
	—	—	—	—	1.76	1.92	2.08	2.26	2.45	2.65
−10	2.60	2.84	3.11	3.38	3.69	4.02	4.37	4.76	5.17	5.62
	2.87	3.10	3.35	3.62	3.91	4.22	4.55	4.90	5.28	5.68
0	6.11	6.57	7.06	7.58	8.14	8.73	9.35	10.02	10.73	11.48
10	12.28	13.13	14.03	14.98	15.99	17.05	18.18	19.38	20.64	21.97
20	23.38	24.87	26.44	28.10	29.84	31.68	33.62	35.66	37.81	40.07
30	42.44	44.94	47.56	50.32	53.21	56.24	59.43	62.77	66.27	69.94

This means that, provided that changes are sufficiently small, because of the non-linearity of the differential expressions, the atmosphere's pressure, after noting that $\mathrm{d}e_s/\mathrm{d}h = \mathrm{d}e_s/\mathrm{d}T \cdot \mathrm{d}T/\mathrm{d}h$, is given by:

$$p_A = \rho g h + e_s + \left(\rho g \frac{\mathrm{d}h}{\mathrm{d}T} + \frac{\mathrm{d}e_s}{\mathrm{d}T}\right) \Bigg/ \left(\frac{1}{V}\frac{\mathrm{d}V}{\mathrm{d}h} \cdot \frac{\mathrm{d}h}{\mathrm{d}T} - \frac{1}{T}\right) \qquad [3.6]$$

Because $\mathrm{d}V/\mathrm{d}h$ and $\mathrm{d}h/\mathrm{d}T$ can readily be determined (be mensuration and noting the sensitivity of the column height to a known temperature change respectively) and standard vapour pressure information such as is summarized in Table 3.1 gives $\mathrm{d}e_s/\mathrm{d}T$, equation [3.6] can readily be used to calibrate the instrument. Provided that the temperature is kept constant, this barometer can be calibrated to be direct reading. The design shown in Fig. 3.1 shows the reference volume enclosed by a water jacket enabling both controlled temperature changes and reduced fluctuations.

An alternative method of preparing the instrument for use is to calibrate it against a suitable standard barometer, so that nR/V may be found from equation [3.3] after inserting measured values of p_A, $\rho g h$, T and e_s.

The sensitivity of this type of barometer is found from equation [3.3] by considering the effect of a small change in atmospheric pressure for a constant temperature T and is given by:

$$\frac{\mathrm{d}h}{\mathrm{d}p_A} = 1 \Bigg/ \left(\rho g - \frac{nR}{V^2} T \frac{\mathrm{d}V}{\mathrm{d}h} + \frac{\mathrm{d}e_s}{\mathrm{d}h}\right) \qquad [3.7]$$

From equation [3.7] it is seen that a liquid column barometer is nearly linear only if the reference volume V is made large and a liquid with a low vapour pressure is chosen. The latter condition can be approached by keeping the barometer at a low temperature.

The limiting factor in the pressure sensitivity of the barometer is the sensitivity of the liquid column to temperature change. Again, from [3.3] it follows that for constant atmospheric pressure:

$$\frac{dh}{dT} = \frac{1}{\rho g}\left[\frac{nR}{V}\left(\frac{T}{V} - 1\right) - \frac{de_s}{dT}\right] \qquad [3.8]$$

A barometer with a reference volume of about 0.75 litre and a column cross-section of about $0.1\,cm^2$ would have a pressure sensitivity of about $1.3\,cm\,mbar^{-1}$ and a temperature sensitivity of $4\,cm\,^\circ C^{-1}$ which figures are indicative of the temperature stability required.

3.3. ANEROID BAROMETERS

Fundamentally this type of barometer consists of an evacuated chamber whose compression by the atmospheric pressure is magnified and indicated by a multiplying system of levers. Temperature corrections are usually necessary due to lever expansion, changes in elasticity of the compressed system and the effect of the small amount of residual gases in the chamber. Many modern designs have been able to compensate for these effects. A common use for the aneroid mechanism is in the barograph, where instead of a simple pointer, a pen arm is caused to move across a scaled chart moved by clockwork.

The great advantages of aneroid barometers are their portability and the fact that they are relatively unaffected by motion. This latter feature makes it possible to use these instruments with a scale of altitude. The altimeter, if accurate, measures the pressure correctly. By setting the zero of the scale to the pressure at sea level, effective altitudes for a standard atmosphere can be read off directly. If the atmosphere is colder or warmer than standard, the instrument will give too high or too low readings of altitude respectively. Normally the readings would not be expected to vary more than 5% from the true values.

Further developments have resulted in an extremely sensitive and accurate form of the aneroid barometer. Sometimes known as the digital aneroid, this device requires no work to be done in moving levers by the moving bellows. A hysteresis-free indication is thus to be expected. A screw micrometer gauge, coupled to a digital counter, is used to measure the bellow's thickness. The reliability of the micrometer measuring system is increased by relying on an electrical indicator to show when the calliper is in the correct position.

These instruments can be housed in a small box, and are thus more portable than mercury barometers. Their sensitivity can be dramatically superior, and their tenure of calibration almost as reliable.

Experiment XIV. The isothermal atmosphere

In this greatly simplified discussion of the earth's atmosphere, the pressure, p, is examined as a function of height, z. Referring to Fig. 3.2, it is seen that if the density of the air at height z is ρ, then:

$$\mathrm{d}p + \rho g \mathrm{d}z = 0 \qquad [3.9]$$

By the ideal gas laws (see equation [3.4]):

$$pv = nRT$$

where v is the volume of n moles of gas at a temperature T. Thus if M is the molecular weight of the gas:

$$p = RT/M \qquad [3.10]$$

it follows therefore from [3.8] and [3.10] that:

$$\frac{\mathrm{d}p}{p} = -\frac{gM}{RT}\mathrm{d}z$$

From which it follows that:

$$p(z) = p(0) \exp\left(-\frac{gM}{RT}z\right) \qquad [3.11]$$

if T is independent of z.

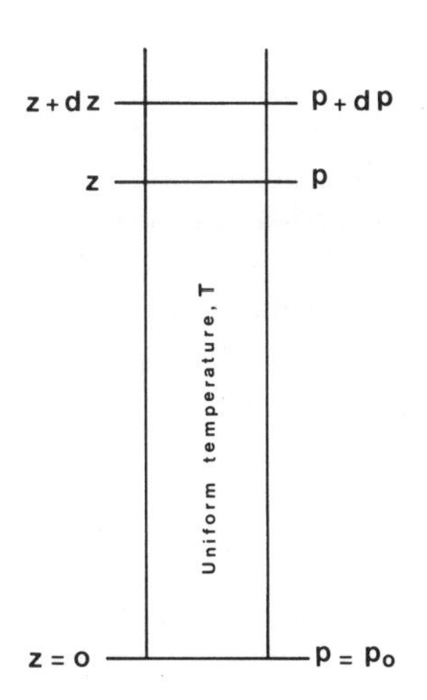

Fig. 3.2. A vertical column through an isothermal atmosphere.

Since z changes much more rapidly than T, equation [3.11] has some validity in a real atmosphere, especially on overcast days when atmospheric temperature gradients may be small.

Over short ranges of height, equation [3.11] is sufficiently accurate to merit experimental consideration. A tall building provides a sufficient pressure variation from the ground floor to the upper storey, which can be measured by any satisfactory apparatus available, preferably a digital aneroid barometer, but the simple water barometer calibrated in Experiment IX would suffice. The air temperature T in equation [3.11] can be taken as the mean over the height range z_1 to z_2, and the expression written as:

$$p(z_2) = p(z_1) \exp \left[-\frac{gM}{RT} (z_2 - z_1) \right] \qquad [3.12]$$

If $p(z_2)$, $p(z_1)$, $(z_2 - z_1)$ and T are measured and g and R are known, then this simple experiment enables the determination of M, the effective molecular weight of the air. Alternatively $(z_2 - z_1)$ can be regarded as unknown after a suitable value of M is introduced. M can be calculated from knowledge of the molecular weights of nitrogen and oxygen.

3.4. ATMOSPHERIC HUMIDITY

Water is supplied to the atmosphere by evaporation from the earth's surface. Thus although the percentage of water vapour in the atmosphere is quite small compared with other gases, this small fraction, because it is continually being transformed and transferred in state and space, feeds the rivers of the earth by precipitation.

The saturation vapour pressure of pure water vapour is defined as the pressure of the vapour when in a state of equilibrium with a plane surface of water at the same temperature. This dependence of the saturation water vapour pressure is a characteristic of the substance, being independent of the pressure of the other gases. The vapour pressure has been determined experimentally over both water and ice. Table 3.1 includes pressures for both ice and super-cooled water at temperatures below freezing. Tables for use with dry- and wet-bulb thermometers embody the vapour pressure with respect to ice at sub-zero temperatures as the wet-bulb thermometer is more reliable if allowed to freeze.

3.5. PARAMETERS SPECIFYING HUMIDITY

Water vapour pressure and density. The equation of state for water vapour also takes the form of the ideal gas equation:

TABLE 3.2

Saturation water vapour density in g m^{-3} as a function of temperature, calculated using the saturation vapour pressure values of Table 3.1 in equation [3.13]. For temperatures below freezing, the upper and lower figures are for ice and supercooled water respectively

Temperature (°C)	Add differences to temperature									
	0	1	2	3	4	5	6	7	8	9
−30	0.34 —	0.38 —	0.42 —	0.46 —	0.50 —	0.55 —	0.61 —	0.67 —	0.73 —	0.81 —
−20	0.88 —	0.97 —	1.06 —	1.16 —	1.27 1.48	1.38 1.61	1.52 1.74	1.66 1.88	1.81 2.04	1.97 2.19
−10	2.14 2.35	2.32 2.54	2.54 2.74	2.75 2.95	2.99 3.17	3.25 3.41	3.52 3.66	3.82 3.93	4.13 4.22	4.47 4.52
0	4.84	5.19	5.56	5.95	6.36	6.80	7.25	7.75	8.27	8.81
10	9.39	10.01	10.66	11.34	12.06	12.81	13.62	14.46	15.35	16.29
20	17.27	18.31	19.67	20.55	21.75	23.01	24.34	25.73	27.19	28.72
30	30.32	32.00	33.75	35.59	37.52	39.52	41.77	43.83	46.12	48.52

$$e = \rho_w \frac{R}{M_w} T \qquad [3.13]$$

where e is the water vapour pressure, M_w the molecular weight of the vapour, T is the absolute temperature and ρ_w is the water vapour density or absolute humidity. It is usually expressed in g m^{-3}. Some typical values of saturated vapour pressure are shown in Table 3.1 and Table 3.2 shows values of the saturated water vapour density. The value of R is 8.314×10^7 erg °K^{-1} mole^{-1} (or 8.314 J °K^{-1} mole^{-1}) and M_w is 18.0 g.

Specific humidity. This quantity is defined as the mass of water vapour contained in unit mass of air (i.e. dry air plus vapour), and is given by:

$$q = m_w/m_a \qquad [3.14]$$

where m_w and m_a are the masses of water and air respectively in a given volume V.

Hence $m_w = \rho_w V$ and $m_a = (\rho_d + \rho_w)V$; therefore, since the subscripts a, d, and w refer to atmospheric air, dry air and water vapour respectively:

$$q = \frac{\rho_w}{\rho_w + \rho_d} \qquad [3.15]$$

where ρ_d is the density of the dry air.

Since $\rho_d = p_d M_d/RT$ and $\rho_w = eM_w/RT$ we have:

$$q = \frac{M_w e}{M_d p_d + M_w e}$$

$$= \frac{M_w}{M_d} \cdot e \Bigg/ \left[p_a - \left(1 - \frac{M_w}{M_d}\right) e \right] \qquad [3.16]$$

where $p_a = p_d + e$, the total pressure.

Using the values $M_w = 18$, $M_d = 28.9$, we have:

$$q = 0.622e/(p_a - 0.378e) \qquad [3.17]$$

This quantity is often expressed in grams of water vapour per gram (or kilogram) air and since e is usually quite small compared with p, equation [3.17] is often written as:

$$q = 0.622e/p_a \qquad [3.18]$$

Mixing ratio. This is defined as the mass of water vapour associated in mixture with unit mass of dry air and is given by:

$$w = \rho_w/\rho_d \qquad [3.19]$$

from which it can be shown that: $w = 0.622e/(p_a - e)$ [3.20]

Since w seldom exceeds $0.02\,g\,g^{-1}$, no appreciable error results on equating specific humidity with the mixing ratio in most meteorological applications.

Relative humidity. For most practical applications this is the temperature-dependent ratio of the actual to the saturated values of any of the absolute humidity parameters.

Dew point temperature. The temperature of the dew point is defined as the temperature at which saturation would just be reached if the air were cooled at constant pressure without removal or addition of water. If the data of Table 3.1 are plotted, the determination of the dew point, provided the actual water vapour pressure is known, is seen to be a simple graphical procedure.

3.6. THE MEASUREMENT OF HUMIDITY

The following list summarizes the various methods of observing humidity:

(1) *Absorption effects.* The absorption of water by suitable materials can be detected as a change in chemical, gravimetric, electrical or other physical properties of this material. Example (2) below is actually only a special case.

(2) *Expansion effects.* Some substances such as hair expand after absorbing moisture. Such changes are readily employed to move a pointer or pen over a scale or moving chart.

(3) *Condensation effects.* Condensation effects depend on the actual observation of the condensation or dew point on a suitably cooled surface.

(4) *Thermodynamic effects.* An example of thermodynamic effects is provided by the different equilibrium temperatures reached by wet and dry surfaces located in the same air stream.

(5) *Electrical properties.* The electrical resistance and dielectric constant of air are dependent on its moisture content.

(6) *Radiation attenuation.* Certain bands of electromagnetic radiation including wavelengths in the infra-red and micro-wave ranges are absorbed by water vapour. Because of their rapid response rate, infra-red "humidiometers" are valuable in the study of turbulent water vapour fluctuations.

Experiment XV. Observation of the dew point

The temperature at which saturation occurs for a given water vapour content is known as the dew point. As supersaturation does not normally occur, this temperature must be lower than or equal to the actual temperature of the air. Thus determination of the dew point allows a direct determination of the specific humidity.

Dew point apparatus may in principle be extremely simple, the temperature of a polished surface being observed at the instant at which dew condenses on it. The chief difficulty in obtaining reliable dew point temperature measurement rests on the difficulty of deciding at just what stage all condensation has disappeared from the mirror surface. In an automated form of this instrument a photo-electric sensor scanning a metallic mirror surface is connected to provide corrective signals to a temperature controlling circuit.

A relatively simple manual method makes use of a thermometer whose sensor has a large surface area to volume ratio. The thermometer is left in a refrigerator to cool to a temperature well below the estimated dew point so that on being brought into the air of the experimental environment, a layer of condensation will form on the thermometer.

In a reverse form of Experiment I, the warming curve is now plotted. Assuming that a relatively constant amount of energy is diverted to vaporizing the condensed moisture, the warming curve will show a change in gradient (on log-linear paper) when all vaporization has ceased. The temperature at this juncture is taken as the dew point, which determination is easily verified by means of an Assmann psychrometer.

Unfortunately, the non-uniformity of the temperatures within the bulbs of thermometers leads to the surface of the sensor having a higher temperature when subjected to external warming than the bulbs of the internal sensor material. This means that the use of mercury-in-glass thermometers, for example, although suitable for illustrating the principle of the experiment, leads to an under-estimation of the dew point temperature. The ideal thermometer for this purpose would be constructed of a thin metal foil on which another metal had been electroplated to create a thermo-junction.

Experiment XVI. The hair hygrometer

Many normally dry organic substances, particularly skin and hair, exhibit changes in dimensions with variations in humidity. The human head-hair increases its length by about 2½% as it comes to equilibrium of air environments from 0 to 100% relative humidity. Although the effect of temperature is not entirely negligible, it is quite small compared to the influence of humidity. This type of sensor is commonly used when a written record is required, as in the hair hygrograph which employs a lever system similar to that used in a mechanical thermograph. When attached to electrical displacement transducers, hair sensors can be used in electrical recording systems.

The response time to changes in humidity is an important factor and can be measured by a technique having some similarity to the methods described in the assessment of thermal inertia. The basis of the method is to bring a hygrometer, after it has come to equilibrium with an environment of known humidity, into a different air environment of the same temperature but changed water vapour content. Basing the observations on humidity indication of the sensor as a function of time, the time constant can be found by treating the humidity parameter as was temperature in Experiment I, i.e.:

$$\tau = t \Big/ \ln \frac{q_0 - q_a}{q - q_a}$$

where the humidities may be measured in any convenient units because the logarithmic expression is of necessity dimensionless. The initial humidity reading of the sensor is q_0 and the values q are those at subsequent times t while q_a is the humidity of the surrounding air.

Controlled air environment enclosures suitable for this type of experimentation can easily be made by gluing together suitably cut sheets of expanded polystyrene foam. Inspection ports can be formed by using two sheets of perspex (a transparent acrylic). Humidity changes can be created by injecting water from a hyperdermic syringe directly through the foam walls.

3.7. DRY- AND WET-BULB THERMOMETRY AND THE PSYCHROMETER

If any ordinary air temperature sensor is covered with a wet muslin cloth it becomes a wet-bulb thermometer. When its reading, T_w, is compared with that of a neighbouring dry thermometer, $T \approx T_a$, it is found that $T_w \leqslant T_a$, the difference depending on the degree of ventilation and the humidity. If the specific humidity is q, and evaporation from the wet-bulb thermometer causes a local humidity of q_w, then $(q_w - q)$ grams of water have evaporated per unit mass of air whose specific heat is c_p. If the latent heat of vaporisation is L, then:

$$(q_w - q)L = c_p(T_a - T_w) \qquad [3.21]$$

therefore:

$$q = q_w - \frac{c_p}{L}(T_a - T_w) \qquad [3.22]$$

Although it might be assumed that $q_w = q_s$ (the saturation value), the simple two-thermometer wet-bulb technique is usually inaccurate because of varying natural ventilation and the effect of radiation, so that even T_a is measured inaccurately.

The instruments known as psychrometers, which exist in many forms, use the simple wet and dry thermometers in systems which incorporate forced ventilation. Standard designs of this type include the sling and whirling psychrometers and the Assmann psychrometer which uses a fan for ventilation.

The balance of thermal energy for a thermometer given in equation [1.8] considered sensible heat flow, H, through the air surrounding the sensor, and net radiation, A_N. This equation must be modified by inclusion of the evaporative heat flux, LE, for application to a wet-bulb thermometer, where L is the latent heat of vaporization and $-E$ is the rate of evaporation, the negative sign arising from the convention of signs adopted. Considering equilibrium to have been reached:

$$H + A_N + LE = 0 \qquad [3.23]$$

where positive signs denote heat gain for the thermometer.

Since (see equation [1.1]):

$$H = \omega(T_a - T_w)$$

$$H + LE = \omega_e(T_a - T_w) \qquad [3.24]$$

where ω_e is a transfer coefficient which holds for wet surfaces and is similar to, but greater than ω, which applies only to dry surfaces, so that:

$$LE = (\omega_e - \omega)(T_a - T_w)$$

$$\therefore LE = (\omega_e - \omega)\frac{L}{c_p}(q_w - q) \qquad [3.25]$$

Using [1.1] and [3.25], equation [3.23] becomes:

$$\omega(T_a - T_w) + A_N + (\omega_e - \omega)\frac{L}{c_p}(q_w - q) = 0$$

and finally:

$$q = q_w - \frac{c_p}{L}\left[(T_a - T_w) + \frac{A_N}{\omega}\right]\frac{\omega}{\omega_e - \omega} \qquad [3.26]$$

TABLE 3.3

Relative humidities (in %) as a function of wet-bulb temperatures (T_w) and wet-bulb depressions below dry-bulb temperatures ($T_a - T_w$)

T_w(°C)	$T_a - T_w$(°C)											
	0.0	0.5	1.0	1.5	2.0	2.5	3.0	4.0	5.0	6.0	8.0	10
0	100	92	83	75	67	61	54	42	31	22	7	0
2	100	92	84	77	70	64	58	47	37	28	14	2
4	100	93	86	79	73	67	61	51	42	33	20	9
6	100	93	87	81	75	69	64	54	46	38	25	15
8	100	94	88	82	76	71	66	57	49	42	29	19
10	100	94	88	83	78	73	69	60	52	45	33	24
12	100	95	89	84	79	75	70	62	55	48	37	28
14	100	95	90	85	81	76	72	74	57	51	40	31
16	100	96	90	86	82	77	74	66	60	54	43	34
18	100	96	91	86	83	78	75	68	62	56	45	37
20	100	97	91	87	83	79	76	69	63	58	48	39
22	100	97	92	87	84	80	77	71	65	59	50	41
24	100	97	92	88	85	81	78	72	66	61	51	43
26	100	98	92	88	85	82	79	73	67	62	53	45
28	100	98	93	89	86	83	80	74	68	63	55	47
30	100	98	93	89	86	84	80	75	69	65	56	48
32	100	98	93	90	87	84	81	76	70	66	57	50
34	100	98	93	90	87	85	82	76	71	67	58	51
36	100	98	94	91	88	85	82	77	72	68	59	52
38	100	98	94	91	88	86	83	78	73	68	60	53

If the rate of ventilation is sufficient, A_N/ω can be made negligible so that [3.26] becomes:

$$q = q_w - \frac{c_p}{L}(T_a - T_w)\frac{\omega}{\omega_e - \omega} \qquad [3.27]$$

The above equation is very similar to [3.22]. At sufficiently high ventilating speeds of about 4 m sec^{-1}, the ratio $\omega/(\omega_e - \omega)$ displays relative constancy, enabling psychrometers to be calibrated.

Table 3.3 shows values of the relative humidity as a function of wet- and dry-bulb temperature data. The relative humidity can be expressed as the ratio of the actual water vapour pressure to the saturation water vapour pressure at a given temperature. As such, the relative humidity does not uniquely specify the water vapour content of the air unless the temperature is also known. Provided both are established then any of the humidity specifying parameters discussed in section 3.5 can be calculated.

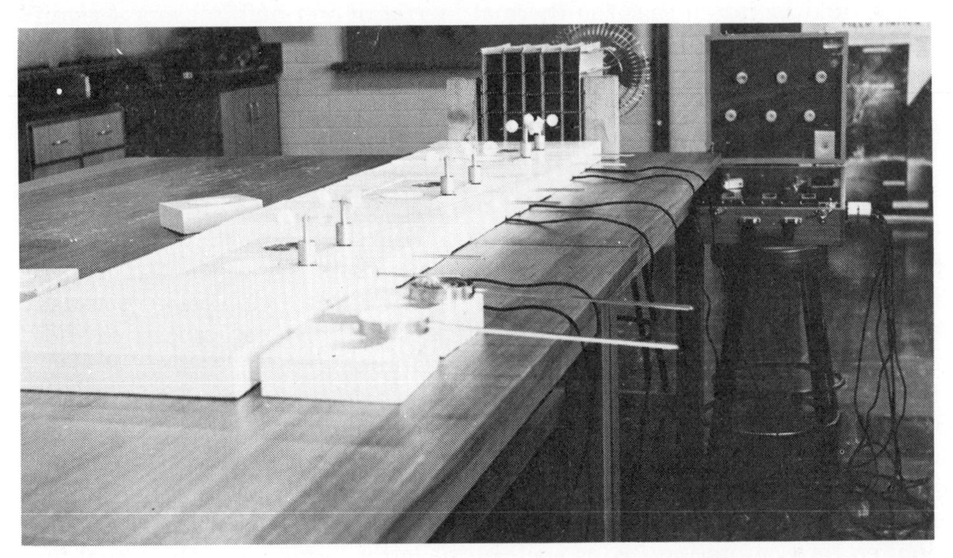

Fig. 3.3 Apparatus illustrating the measurement of the Bowen ratio. Polyurethane foam blocks insulate all except the upper surfaces of solid and annularly hollowed-out aluminium disks. The latter is filled with water, rubber grommets preventing leaks at the point of thermometer insertion. The foreground shows the two types of blocks with thermometers, removed from the closely fitting depressions in the foam blocks normally protecting them. Above each surface, small cup anemometers record the speed of the air stream issuing from the cardboard collimator in front of an electrical fan. The case on the right houses the electro-mechanical counters registering anemometer rotations.

Experiment XVII. The ventilated wet-bulb thermometer

In an experimental arrangement similar to that employed in Experiment II, an electrical blower is set up to direct its draught along a table so that several small groups of observers can sit alongside to read thermometers supported in air of varying speed. It is important, of course, to co-operate in keeping interference with the flow of air to the downstream observers to a minimum. The air speed at each relevant point can be established with a small air meter or small anemometers, as in Experiment II.

At each observational point, several of which should be selected to provide a representative range of air speeds, two thermometers are supported in the draught, one of which in each case should be equipped with a wet wick. The observations simply consist of noting the wet-bulb depression as a function of ventilating air speed so as to establish the minimum air speed requirements for satisfactory psychrometer operation.

If the value of $(q_w - q)$ in equation [3.27] is established independently, for example by means of a calibrated psychrometer, then the value of the factor $\omega/(\omega_e - \omega)$ can be found as a function of ventilating speed, simply by observing the value of $(T_a - T_w)$.

It should be noted that the thermal inertia of wet-bulb sensors is significantly greater than for their dry-bulb counterparts unless the latter are selected to be of greater physical dimensions.

Experiment XVIII. Measurement of the Bowen ratio

In micro-meteorology, the Bowen ratio is an important practical concept which expresses the ratio of the energy involved in sensible heating of the air and in evaporation of water at the lower boundary of the atmosphere. Over natural surfaces which include vegetation, the evaporation is usually termed evapo-transpiration in recognition that the total process includes physical evaporation and biological transpiration.

Experiment II has shown how an empirical heat transfer coefficient ω can be determined for a dry surface. If this experimental is applied to a wet surface, then a similar coefficient ω_e for this type of surface can be determined. In this instance, heat is transferred not only by sensible heating H and net radiation A_N, as indicated in equation [1.11], but the latent heat required for evaporation LE, as well. In the latter term, $-E$ is the actual flux of water vapour and L is the latent heat of vaporization. If the net radiation term is the same in both instances, which for surfaces both wet and dry at similar temperatures in similar room temperature environments free from varying levels of intense visible radiation is a reasonable assumption, then:

$$\omega(T_a - T) = H$$

and:

$$\omega_e(T_a - T) = H + LE \qquad [3.28]$$

$$\frac{H}{LE} = \frac{\omega}{\omega_e - \omega}$$

Equation [3.28] is an expression for the Bowen ratio.

The experiment relies partly on the results of Experiment II and continues with the use of a modified form of the apparatus. Hollowed-out metal disks with annular fins as shown in Fig. 3.3 allow a water surface to be exposed to an air stream whose velocity is measured as before, either by an air meter or by suitable miniature sensitive anemometers. Initially hot water is used to fill the hollows in the disks and the cooling rate is observed by regular temperature measurement. The rate of loss of heat, which is assumed to be entirely through the upper surface, the others being insulated, can then be calculated if the masses of water and metal are found by suitable weighing procedures.

CHAPTER 4

Wind velocity and turbulent transfer

The velocity of the wind is a vector quantity and consequently requires the specification of both speed and direction for a given time and location. While some micro-meteorological applications call for measurements of wind parameters with response times of a fraction of a second, synoptic studies are adequately served by mean values over periods of time of the order of an hour. Clearly there might be marked differences between the sensors used in each of these extreme cases. The former requires anemometers of low inertia while the latter has data reduction simplified if the anemometer employed does not respond to high-frequency fluctuations but effectively to the mean wind speed, by virtue of its inertia.

4.1. METHODS OF WIND SPEED MEASUREMENT

The instruments used for measuring wind speed can be classed into five main groups.

(1) *Rotation anemometers.* These devices include both cup and vane rotors on vertical axes and propellers. The response of a propeller depends on its orientation with respect to the direction of the air flow. If the flow is uniform in space and time appropriate orientation of the sensor can be achieved manually. Such conditions are to be expected only in wind tunnels. In the atmosphere the mean wind flow is the resultant of eddies of various sizes so that at any real location the direction and speed of air varies continually. Anemometers with a vertically mounted rotor of cups are insensitive to variations in horizontal direction of air flow. Large sensors are able to integrate out the smaller localized fluctuations in space, while an increase in mass or inertia will smooth out variations in time.

Thus anemometers required to monitor the mean air flow characteristics of a general location should be massive and sample over a large cross-section of the flow.

Propeller anemometers if used singly are of necessity equipped with a locating vane.

Correctly designed light-weight propellers can have a near cosine response to wind direction and may therefore be used to measure vector components of the wind. If the three orthogonal components of the wind velocity are given by u, v and w, there are micro-meteorological applications calling for instantaneous measurements of respectively these three components or alternatively $\sqrt{u^2 + v^2}$ and w. The latter set is obtained by having one propeller on a vane rotating in the horizontal plane and the third fixed to monitor only vertical air movement. In order to achieve lightness of weight, such sensors are often constructed of expanded polystyrene.

Rotor or cup anemometers are usually designed to monitor characteristics of horizontal air flow only. If information on the direction is required, a separate vane is necessary.

The response of a wind vane depends on its inertia, area and geometry, the choice depending on whether the requirement is to observe turbulent or mean flow characteristics. A reduction in the sensitivity to small eddies can be achieved by means of mechanical or hydrodynamical damping as well as by built-in massive inertia.

Where a fast response time and slow starting speed are important, the output from anemometers may be obtained by the frictionless chopping of a beam of light or by magnetic proximity switching. One interesting method of reducing the effective starting speed of cup anemometers is to rotate the entire body constantly at the normal minimum starting speed. The only inconvenience is the need to add this value to all observations.

Cup anemometers are often used in micro-meteorological profile measurements of the atmosphere boundary layer. In this application slow starting speeds are essential.

(2) *Pressure plate anemometers.* This group of wind sensors can be divided into the swinging type and normal incidence pressure plate anemometers. The former involve a hinged plate which the force of the wind causes to swing and either balance the force of gravity or that of a spring. In both of these cases the cross-sectional area presented by the plate to the wind changes with wind speed. The normal incidence plate is constrained not to change its orientation with respect to the wind direction and the force against the constant area presented can be measured in a variety of ways including the compression of a spring, the movement of a weight on a lever, the displacement of liquid from a piston or the compression of a gas and to give an electrical example, to make use of a strain gauge.

If the velocity of air of density ρ is V, then the force on a normally held flat plate of area A is given by:

$$F = cA\rho \frac{V^2}{2} \qquad [4.1]$$

where $c \approx 1$, but varies somewhat with plate shape and size.

(3) *Pressure tube anemometers.* This type is related to the previously discussed pressure plate sensors. Usually, however, a refinement introduced in that the pressure difference is observed between two chambers whose

openings are designed to be respectively normal and parallel to the air flow. Such devices are known as Pitot tubes. The air pressure difference between the two chambers is related to the wind velocity V and density of the air ρ, by Bernoulli's equation:

$$\Delta p = \tfrac{1}{2}\rho V^2 \qquad [4.2]$$

$$\therefore V = \sqrt{2\Delta p/\rho} \qquad [4.3]$$

Equation [4.2] is, however, not completely accurate and should incorporate a constant similar to that introduced in equation [4.1].

(4) *Thermal anemometers.* Two main types of thermal wind speed sensor can be devised. The first is exemplified by the approach that involves measuring the rate of cooling of a heated thermometer. As has been shown in Experiment II, the rate of cooling, if expressed in suitable non-dimensional form, is a monotonic function of the wind speed. This type of measurement clearly results in integration over the period of time required for the observation of cooling so that the method cannot be used for continuous monitoring.

The second group of sensors include the hot wire and thermistor-bead anemometers, although a variety of transducers may be used. All such anemometers depend on having a sensor heated by a controlled or constant source of power. Under these conditions, the temperature of the sensor is dependent only on the temperature of the surroundings and on the effective ventilation. The temperature can be monitored either directly by measuring the resistance of the conductor or indirectly by means of an adhered thermocouple junction, for example.

At low power inputs, the temperature and hence the resistance depends largely on the temperature of the surroundings in the manner of a resistance thermometer. With increasingly high levels of power input, the equilibrium temperature, determined by the rate of heat loss, progressively becomes controlled by the ventilation or wind speed. An unsatisfactory dual dependence on both air temperature and speed occurs at intermediate rates of heating.

(5) *Acoustic anemometers.* Acoustic or sonic anemometers rely on the dependence of the transit time of a sonic pulse on the transmitting medium, in this case, air. Although allowance must also be made for temperature fluctuations, these instruments, in avoiding inertial sensing components, are able to be used in the study of high-frequency eddies.

Experiment XIX. The comparison of anemometers

The comparison of a variety of anemometers is best achieved in a uniform air stream, a requirement that makes natural winds far from ideal. For many purposes, a satisfactory wind tunnel can be built from sheets of expanded polystyrene glued together to form a cylinder of square cross-sectional area large compared to that presented to the air flow by anemometers to be used. A large electric fan, whose speed can be adjusted by means of an autotransformer is mounted at one end and arranged to blow through grill into the elongated experimental chamber which can be observed through perspex windows press-fitted into holes cut into the walls.

For the subsequently described wind profile experiment, a set of four or more cup anemometers, preferably designed for micro-meteorological purposes, is essential. These can be usefully intercompared in the tunnel described. If a standard calibration is available for a mechanical rotor anemometer it can be used in comparison with other types of anemometer.

A swinging pressure plate anemometer can be constructed from simple materials. A fixed normal incidence plate equipped with a strain gauge can be calibrated for force indication in a variety of ways. However, this must be done without changing the orientation of the plate from that intended during actual operation. Thus the force from a standard weight must be translated horizontally by means of a lever.

The main problem encountered with hot wire anemometers in the simple test enclosure described results from the lack of sufficient thermal inertia to damp out the effect of turbulent velocity fluctuations unless a relatively thick wire is employed. Similarly if the measurements are being performed with a thermiston sensor, a relatively large semi-conductor should be selected. In most applications, the rapid response rate of lightweight forms of these thermal anemometers is the main attribute for which they are selected. Such sensitive devices require a more elaborate calibration facility, however.

If a wire with a suitably large temperature coefficient of resistance, such as nickel, is used, the measured electrical resistance for either a constant current or a constant applied potential difference can be empirically related to the wind speed. The temperature, T, of the wire must be sufficiently high that the greatest anticipated magnitude of variations $|\Delta T_a|$, in the air of mean temperature $\bar{T}_a$, is insignificant when compared to $(T - \bar{T}_a)$. Since the heat transfer coefficient, ω, has been determined in Experiment I for a variety of surface shapes, the heat loss from the electrically heated wire, as calculated from equation [1.1] may be compared with the measured rate of dissipation of electrical power in the resistance wire, as in Experiment IV.

A Pitot tube anemometer can also be built from basic materials such as brass tubing in which clean cut holes can be drilled. The two tubes constituting the separate pressure chambers need not be coaxial but may be parallel-mounted tubes of the same diameter, one with a hole in its forward end, the other with a hole along its side wall. Because of its relatively small cross-sectional sampling area, this wind sensor can respond to the turbulent fluctuations making up the mean air flow. Its response time depends principally on the inertia of the volume of displaced liquid in the associated manometer as well as its viscosity. The mass of liquid displaced in a given manometer U-tube depends only on the pressure difference applied and not on the density of the indicating liquid. Hence the frequency response must be controlled by means of selecting a liquid of suitable viscosity. For this purpose a variety of light oils should be available.

This anemometer can be usefully employed in determining the relative variations in wind speed encountered within the experimental volume of the wind tunnel without requiring knowledge of the empirical correction factor.

4.2. THE WIND VELOCITY PROFILE IN THE ATMOSPHERIC BOUNDARY LAYER

Horizontal momentum is transferred from higher levels in the moving atmosphere toward the ground where it is dissipated in frictional drag as evidenced by the observation that very close to the ground, the velocity becomes zero.

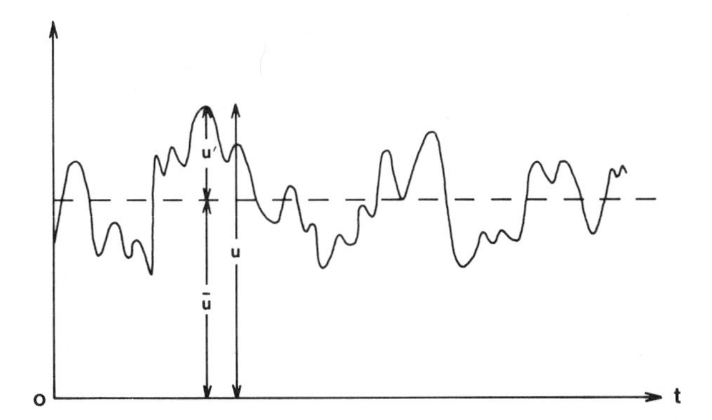

Fig. 4.1. A graphical illustration of the relationship between instantaneous and mean values of velocity as well as fluctuations.

If the wind is monitored by anemometers of sufficiently rapid response, under most conditions the natural air stream is found to fluctuate markedly about a mean value and further, if the velocity is specified in terms of three orthogonal components, u, v, and w, fluctuations u', v' and w' about mean values $\bar{u}$, $\bar{v}$ and $\bar{w}$ respectively are observed as shown in Fig. 4.1. While the relative magnitude of the mean horizontal components $\bar{u}$ and $\bar{v}$ depends on the direction of the axes to the mean wind direction and in fact could be selected so that $\bar{v}$, for example, becomes zero; $\bar{w}$ is always zero under neutral atmospheric conditions. This is because in spite of eddies having both upward and downward components, there is no net upward or downward movement of air over any longer period of time when the atmosphere is neutral in stability.

The horizontal momentum of unit volume of air at a given level at a given instant of time is ρu. If this same volume of air also has an instantaneous vertical velocity w, then the downward transfer of horizontal momentum at that instant is given by:

$$\tau = -\rho u w \qquad [4.4]$$

It follows that the mean downward transfer of momentum is:

$$\bar{\tau} = -\overline{\rho u w} \qquad [4.5]$$

provided that ρ, the density of the air, is constant.

Since $u = \bar{u} + u'$ and $w = \bar{w} + w' = w'$, because $\bar{w} = 0$, it follows that:

$$\bar{\tau} = -\rho\overline{(\bar{u} + u')w'}$$

$$= -\overline{\rho\bar{u}w' + \rho u'w'} = -\rho\bar{u}\bar{w}' + \rho\overline{u'w'}$$

$$= -\rho\overline{u'w'} \qquad [4.6]$$

From equation [4.6] it is found that the ratio $\sqrt{\tau/\rho}$ has the dimensions of velocity and is in fact written as u_*, the friction velocity, a quantity that can also be regarded as having the magnitude of the geometric mean of the horizontal and vertical velocity fluctuations.

The observation of these fluctuations requires sensitive low-inertial anemometers. The effect of these fluctuations on vertical transfer of horizontal momentum can also be studied by considering the characteristics of the mean flow at various levels.

Suppose that the vertical gradient of mean horizontal velocity is $\partial\bar{u}/\partial z$. Then if the mean horizontal velocity at a height z is $\bar{u}$, it follows that at a height $z + l$, the velocity becomes $\bar{u} + (\partial\bar{u}/\partial z)l$. Considering now a volume of air initially moving at level z at a speed $\bar{u}$, suddenly swept to the height $z + l$ as an eddy with unchanged horizontal velocity. This parcel of air can now be regarded as injecting a fluctuation in air speed because at the new height $z + l$, the mean velocity is normally $\bar{u} + (\partial\bar{u}/\partial z)l$. The fluctuation in momentum can thus be written:

$$\rho|u'| = \rho\left|\frac{\partial\bar{u}}{\partial z}\right| l \qquad [4.7]$$

Because a fundamental observation of isotropically turbulent flow is that $\overline{u'^2} = \overline{w'^2}$, the mean vertical transfer of horizontal momentum now follows as:

$$\bar{\tau} = -\rho\overline{u'w'}$$
$$= -\rho \left(\frac{\partial \bar{u}}{\partial z}\right)^2 l^2 \qquad [4.8]$$

Equation [4.8] is rewritten as:

$$\bar{\tau} = -\rho k^2 z^2 \left(\frac{\partial \bar{u}}{\partial z}\right)^2 \qquad [4.9]$$

where k, von Karman's constant has a dimensionless value of approximately 0.4.

Substituting the expression for the friction velocity it follows that:

$$\frac{\partial \bar{u}}{\partial z} = \frac{u_*}{kz} \qquad [4.10]$$

Integration of equation [4.10] leads to:

$$\bar{u} = \frac{u_*}{k} \ln \frac{z}{z_0} \qquad [4.11]$$

where z_0 is the height above the ground where the velocity of the air vanishes and is known as the roughness length, a characteristic of the surface.

For smooth tarmac surfaces, roughness lengths of approximately 0.002 cm might be expected while for grass of length from 2 to 10 cm this value would lie between 0.1 and 1 cm. Because the geometry of flexible surfaces such as presented by grass depends on the wind speed, increase in the latter tends to decrease the roughness length as the vegetation becomes flattened. Over water, on the other hand, the height of developed waves increases with wind speed. Thus the aerodynamic roughness of the surfaces of oceans and lakes might generally be expected to increase with wind velocity.

Experiment XX. Observation of the mean wind profile

The theoretical expressions that have been derived above are strictly valid only under conditions of neutral stability. A windy overcast day would provide conditions approximating this requirement. It is most important that the notes for Experiment XXI be consulted to help determine the general suitability of available sites.

A vertical array of at least four micro-meteorological cup anemometers mounted on a slim mast at heights ranging up to 4 m should be set up and a number of determinations of wind speeds made over 5- or 10-minute periods. A suitable arrangement is shown in Fig. 4.2.

The value of $\bar{u}$ for each corresponding value of z should be plotted on semi-logarithmic graph paper on which these points should lie on an approximately straight line whose extrapolation to $\bar{u} = 0$ allows the roughness length z_0 to be read off directly.

Fig. 4.2. Anemometers mounted at logarithmically progressing heights: 10, 20, 40, 80, 160 and 320 cm above the ground. Downwind, an observer is shown reading the electromechanical counters which record the anemometers' rotations.

Given that von Karman's constant, $k \approx 0.4$, u_* can now be calculated by substitution of observed values of $\bar{u}$, z and z_0 in equation [4.11]. If Experiment XXII is performed simultaneously, the value of $\overline{u'w'}$ can be obtained independently and compared with the value of u_* found here, thus providing a fundamental check of the validity of the mean profile theory and the precision of the separate experimental methods.

An interesting extension of the profile experiment involves artificially modifying the surface of the ground by setting out a regular array of large household buckets upwind from the anemometer mast. At least 100 buckets are required. The observed velocity profile should again be plotted and the values of z_0 and u_* be determined as before. It will be observed that the magnitude of z_0 bears no direct relationship to the height of the obstructions or individual roughness elements because of the effect of variable horizontal spacing.

Experiment XXI. The effect of obstructions on the wind profile

One of the major practical problems in micro-meteorological investigations involves the location of sites over which the wind profile is not unduly perturbed by obstructions to the flow which are atypical of the surface under investigation. As a general working rule for careful measurements, the height of obstructions above the horizon, of substantial horizontal extent, are preferred to be less than 1/100 of their distance from the profile mast. Although such relatively ideal circumstances are rarely encountered in an urban environment, it is clearly quite impossible to obtain useful data in the vicinity of buildings and stands of tall trees. Large, unenclosed sporting grounds may provide reasonable conditions in certain circumstances.

Assuming that an area has been located where a fair logarithmic wind profile can be observed, artificial obstructions constructed of timber or fencing materials, for example, can be introduced to provide various height-to-distance ratios. This should be done by separately altering both height and distance of the obstruction or "wind-break" from the mast.

Experiment XXII. Determination of momentum transfer by the eddy correlation method

If suitable lightweight propeller anemometers are available, simultaneous values of u and w can be recorded. The preferred horizontal sensor is one kept oriented into the wind direction by a lightweight vane.

A two-channel continuous line recorder is required to provide simultaneous traces of the output from the orthogonal sensors for some minutes. The manual process of subsequently calculating $\overline{u'w'}$ is laborious but quite feasible and best done by digitizing the records at suitable, equal time intervals.

The observations should be carried out with the sensors at two different heights, say 2 and 8 m respectively, preferably simultaneously. Because the magnitude of the eddies increases with height, the differences in the value of $\overline{u'w'}$ determined at the two levels will be indicative of the significance of higher frequencies escaping detection in the system used. If performed simultaneously to Experiment XX, the values of $u_* = \overline{u'w'}$ can be compared directly.

Provided that the sampling distance is not too great, inertialess acoustic anemometers designed to measure the separate orthogonal components of air velocity would be ideal. The fidelity of response of propellers to turbulent fluctuations depends not only on their upper limits of frequency response, determined by mechanical inertia, but also on their cross-sectional sampling area whose magnitude may mask the passage of smaller eddies by spatial integration.

Fig. 4.3. Gill-designed vertical (right) and horizontal (left) propeller anemometers, the latter equipped with a locating vane. Each of the lightweight polystyrene foam propellers drives a small direct current generator whose output is a calibrated function of the wind speed. The vertical propeller has an upper extension shaft so that both upward and downward air flows face similar impedances.

The photograph (Fig. 4.3) shows a functional arrangement of vertical and horizontal propeller anemometers. Such devices are relatively expensive, so that the alternative instrumentation depicted in Fig. 4.4 may be desirable in some circumstances. In this latter arrangement, the mean horizontal wind speed is measured by means of a simple cup anemometer, thereby yielding $U = \sqrt{u^2 + v^2} \approx \sqrt{u^2 + v^2 + w^2}$ since $w^2 \ll u^2 + v^2$. The biaxial bivane shown in the photograph can be constructed from suitable lightweight materials, using two ball-bearing shafted continuous potentiometers as transducers for the spatial orientation of the attached vane.

The calibrated outputs from the potentiometers will yield instantaneous values of the altitude θ and azimuth ϕ of the wind vector. Since $\bar{\theta} = 0$,

$\theta' = \theta$, (the meaning of the prime symbol being as before), so that:

$$w' = \sqrt{u^2 + v^2 + w^2}\,\sin\theta \qquad [4.12]$$

Similarly, if the mean horizontal wind direction is chosen to prescribe that $\phi = 0$ and hence $\phi' = \phi$, then:

$$u' = \sqrt{u^2 + v^2 + w^2}\,\sin\phi \qquad [4.13]$$

and hence:

$$u'w' = (u^2 + v^2 + w^2)\,\sin\phi\,\sin\theta$$
$$= U^2 \sin\phi\,\sin\theta \qquad [4.14]$$

Fig. 4.4. A simple bi-axial vane used in conjunction with a horizontal cup anemometer. The vane must be carefully balanced in still air before used in an experiment. The altitude angle-transducing potentiometer is clearly seen whilst the lower housing conceals the similar azimuth transducer as well as the slip rings transmitting the signal from the upper unit.

4.3. THE SCALE OF TURBULENCE

The mean vertical flux of horizontal momentum τ or alternatively the function velocity u_* related by (see equation [4.6]):

$$\tau = -\rho\overline{u'w'}$$

are both parameters that clearly provide some specification of the magnitude of turbulent fluctuations. If the respective variations u' and w' in the mean horizontal and vertical velocities $\bar{u}$ and $\bar{w}$ are in phase, then the friction velocity u_* is proportional to the amplitudes of u' and w'.

The quantity defined as the cross-correlation between u' and w':

$$R_{u'w'} = \overline{u'w'} \Big/ \sqrt{\overline{u'^2} \cdot \overline{w'^2}} \qquad [4.15]$$

is a non-dimensional parameter indicative of the nature of the turbulent eddies. The maximum value of one occurs when u' and w' are completely in phase and the minimum of minus one results when these fluctuations are completely out of phase.

A measure of the time scale of turbulent fluctuations is given by the auto-correlation of the fluctuations of a given velocity component. The auto-correlation is defined as the correlation between the value of the parameter at some time t_1 and at a later time $(t_1 + t')$:

$$R_{u'}(t') = \overline{u'(t_1) \cdot u'(t_1 + t')} \Big/ \overline{u'^2(t_1)} \qquad [4.16]$$

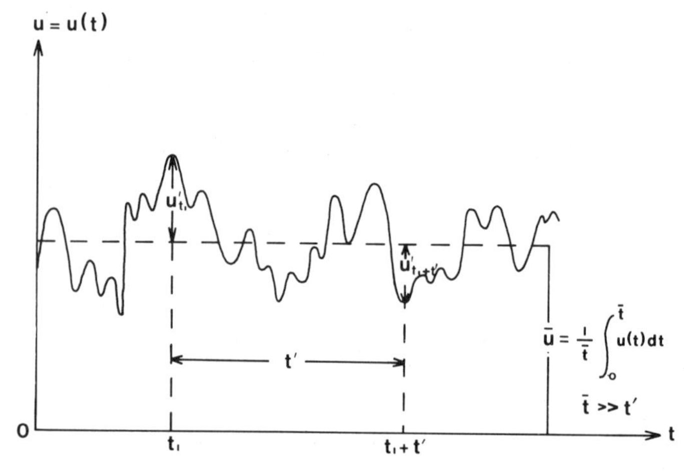

Fig. 4.5. Graphical representation of the variables involved in auto-correlation calculations.

If the nature of the turbulence remains unchanged during the period of time $\bar{t}$ over which the mean is taken, then the actual value of the reference time t_1 for the auto-correlation is unimportant. Invariably $R_{u'}(0) = 1$. Fig. 4.5 shows the relationship between the times t_1, t' and $\bar{t}$.

The behaviour of the auto-correlation as a function of t depends strongly on the time scale of the turbulent fluctuations. For example, if $u'(t_1 + t')$ varies only very slowly (for fixed t_1 and variable t) then $R_{u'}(t')$ falls only slowly from its maximum value of $R_{u'}(0) = 1$ in a gradual approach to zero. On the other hand, the auto-correlation for rapid velocity fluctuations is characterized by a correspondingly rapid fall to zero. In all cases some statistical "wobble" about the zero would be expected for the value of $R_{u'}(t')$ for higher values of t'.

Although the auto-correlation is dimensionless, it is a function of the elapsed time t. Thus in order to derive a single parameter to describe the time scale of turbulence the time t_e, taken for the value of $R_{u'}(t)$ to fall from 1 to $1/e$ can be introduced:

$$R_{u'}(t_e) = 1/e \qquad [4.17]$$

In general t_e is most easily determined from a graphical plot of $R_{u'}(t')$.

Experiment XXIII. The time scale of turbulent fluctuations

The low-inertia anemometers used in Experiment XXII should again be employed. If two anemometers are available, one should be kept at a height of 2 m to monitor the horizontal wind velocity and its fluctuations, the other should be operated successively at heights of 4 and 8 m. Mean velocities should be obtained over 10-minute intervals and the auto-correlations calculated for the three heights. In the data analysis, the time increments chosen should be small enough to avoid masking the frequency components recorded to any significant extent. The value of the time t_e as defined in equation [4.17] should be plotted as a function of height.

4.4. TURBULENT TRANSFER

In section 4.2 the mean vertical transfer of horizontal momentum has been discussed. The same argument that showed that (see equation [4.6]):

$$\bar{\tau} = -\rho\overline{u'w'}$$

can readily be employed to show that the mean vertical transfer of heat and water vapour are respectively given by:

$$\bar{H} = \rho c_p \overline{T'w'} \qquad [4.18]$$

and:

$$\bar{E} = \rho\overline{w'q'} \qquad [4.19]$$

where T' and q' are respectively fluctuations in air temperature and specific humidity and c_p is the specific heat at constant pressure. By including the latent heat of vaporization, equation [4.19] is readily converted to express the energy flux of vaporization:

$$L\bar{E} = \rho L\overline{w'q'} \qquad [4.20]$$

The negative sign on the right-hand side of equation [4.6] is absent from [4.18] and [4.19] as horizontal momentum is transferred downwards (to the earth's surface) whilst the fluxes of sensible heat and water vapour are directed upwards, having their sources at the earth's surface.

Although the use of equation [4.6] in the "eddy correlation" technique is within the scope of relatively basic experimental observations, the same cannot be said about this method for the determination of sensible heat transfer and evaporation through equations [4.18] and [4.19] or [4.20].

An alternative approach introduces analogies between molecular (kinetic) and turbulent transfer processes and simply introduces the following equations:

$$\bar{\tau} = -\rho K_{\mathrm{m}} \frac{\partial u}{\partial z} \qquad [4.21]$$

$$\bar{H} = \rho c_p K_{\mathrm{H}} \left(\frac{\partial \bar{T}}{\partial z} - \Gamma\right)$$

$$= \rho c_p K_{\mathrm{H}} \frac{\partial \bar{\theta}}{\partial z} \qquad [4.22]$$

$$\bar{E} = \rho K_{\mathrm{W}} \frac{\partial \bar{q}}{\partial z} \qquad [4.23]$$

where K_{m}, K_{H} and K_{W} are turbulent transfer coefficients for horizontal momentum, sensible heat and water vapour respectively, Γ is the adiabatic lapse rate and $\bar{\theta}$ is the potential temperature. Since these transfer coefficients must be related to the scale of the turbulent eddies, evidence of Experiments XX and XXII will have shown that all of these coefficients must be expected to increase with height. Experiment XXII shows the variation in scale of turbulence with height and Experiment XX quite clearly demonstrates the fact that the gradient of the mean horizontal velocity, $\partial\bar{u}/\partial z$, falls off with height, a basic assumption of micro-meteorological measurements, then K_{m} must increase accordingly.

Since (see equation [4.9]):

$$\bar{\tau} = -\rho k^2 z^2 \left(\frac{\partial \bar{u}}{\partial z}\right)^2$$

and (equation [4.10]):

$$\frac{\partial \bar{u}}{\partial z} = \frac{u_*}{kz}$$

it follows that:

$$\bar{\tau} = -\rho u_* kz \frac{\partial \bar{u}}{\partial z} \qquad [4.24]$$

Comparison of [4.24] and [4.21] shows that:

$$K_m = u_* kz \qquad [4.25]$$

In order to avoid K_m vanishing at $z = 0$, equation [4.25] is usually modified to include the roughness parameter:

$$K_m = u_* k(z + z_0) \qquad [4.26]$$

showing that the magnitude of this transfer coefficient increases linearly with height.

Experiment XXIV. Turbulent transfer of heat and water vapour

Because of the experimental difficulties in making use of the equations [4.18] and [4.19] it is simpler to design an experiment to measure the mean fluxes of sensible heat and water vapour in the atmosphere on equations [4.22] and [4.23]. Since heat and water vapour have a common source in most situations, it is reasonable to assume that a common transfer mechanism would lead to $K_H = K_W$. Although the transfer coefficient for horizontal momentum K_m cannot be included in this equalizing assumption a basic estimate of the heat and water vapour fluxes can be made by assuming that $K_m \approx K_H = K_W$. On this basis, K_m (or the general transfer coefficient) can be found from equation [4.26] if u_* and z_0 are determined from the vertical profile of mean horizontal velocity.

For the determination of the vertical heat flux then, the vertical distribution of both horizontal velocity and temperature is required while the water vapour flux necessitates the measurement of the vertical gradient of wet-bulb temperatures (or other sufficiently sensitive humidity parameter) as well.

Wind measurements should follow the method of Experiment XX. Temperatures must be established to within better than 0.05°C, a requirement that is usually met by a suitable choice of resistance thermometers or thermistors. Since mean values are required and fluctuations unwanted, small thermistors should be substantially thermally lagged.

An alternative approach also makes use of equations [4.22] and [4.23] and makes no assumptions with respect to K_H and K_W beyond their equality. It thus follows that for a finite height interval:

$$\frac{\bar{E}}{\bar{H}} \approx \frac{1}{c_p} \frac{\Delta \bar{q}}{\Delta \bar{\theta}} \qquad [4.27]$$

The method relies on an independent measurement of either $\bar{E}$ (or $\bar{H}$) or $\bar{E} + \bar{H}$ and although specialized instruments, such as lysimeters, suitable for the measurement of $\bar{E}$ in many situations exist, it is more instructive to obtain the information from the terrestrial heat balance equation:

$$A_N = G + H + LE \quad [4.28]$$

Fig. 4.6 illustrates the sign convention adopted, in which net radiant energy flow, A_N, to the terrestrial surface is regarded as positive as are the conducted heat flow into the ground, G, sensible heat flux, H, and evaporative heat flux, LE, into the atmosphere from the surface.

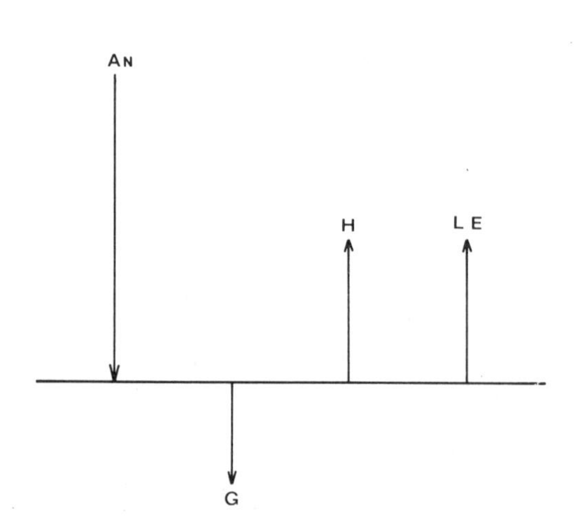

Fig. 4.6. Meteorological energy fluxes above and below the surface of opaque, solid ground

If A_N can be found by using a net radiometer and the value of G obtained (by one or more of the methods described in Chapter 5) then the sum of the atmospheric fluxes $H + LE$ can be found by difference in equation [4.28]. Thus if the ratio and the sum of H and LE are known, then the individual values can be calculated, The ratio referred to follows, of course, from equation [4.27].

In this type of experiment, temperature measurements require great care. Reference is made to section 1.6 in which a suitable shield for temperature sensors (including resistance thermometers and thermistors) has been described. Fig. 1.6 shows the significant constructional details of this shield and Fig. 4.7 shows a photograph of both the basic device as well as a similar unit modified by the provision of a controlled water supply allowing for the continuous monitoring of wet-bulb temperatures. The water flow is controlled by a simple automotive carburetter float-control valve which ensures a constant water level in an adjacent small

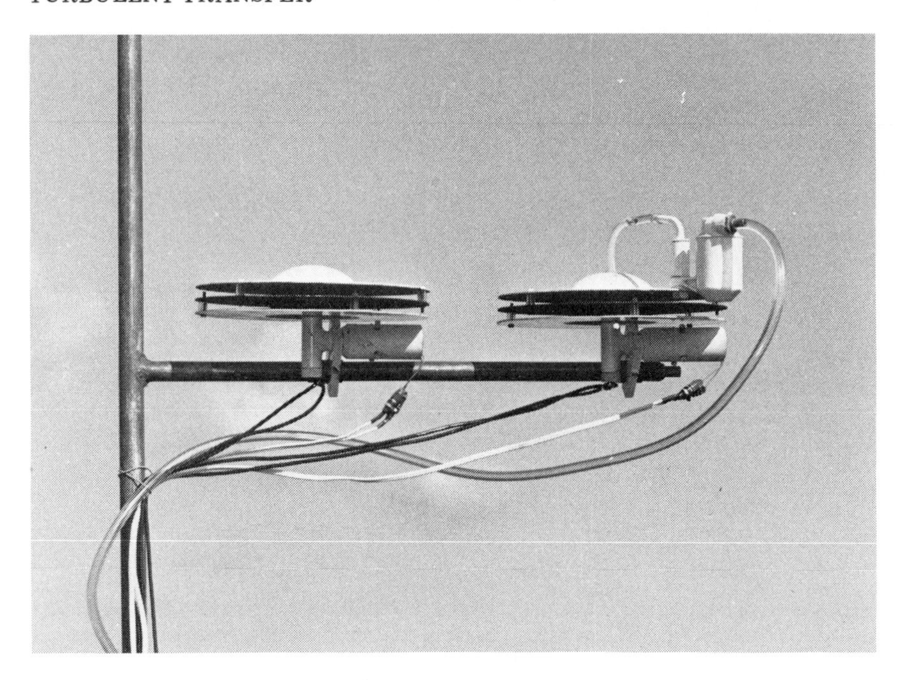

Fig. 4.7. Ventilated shields for dry- and wet-bulb electrical thermometers, with the former being on the left in the photograph. On the wet-bulb unit (right), the float-control valve chamber, smaller constant water level reservoir and the protection tube for the wick are clearly shown (see also Fig. 1.6).

reservoir to which the wet-bulb wick is connected via a protecting tube.

A fundamentally alternative experimental method of assessing the surface heat balance employs a lysimeter to yield the actual value of E and hence LE directly, so that if A_N and G are obtained independently as well, the value of H follows by difference in equation [4.28]. Moreover, if the atmospheric turbulent flux measurements are now obtained an overall assessment of the data on surface energy partition becomes available.

A lysimeter involves the creation of a facility for weighing a representative sample of the material including, and below, the terrestrial surface. This clearly rules out the use of these devices for short-term or mobile observations. Nevertheless, for long-term micro-climatological work lysimeters can be invaluable for monitoring evaporation and plant transpiration, usually referred to in combination as evapo-transpiration. Precise lysimetry necessitates an expensively engineered installation. However, a simpler adaptation of the method, based on the regular weighing (by means for example of a butchers' carcase beam balance) of a ground or soil sample contained by a cylindrical steel drum, normally sited in a closely jacketting hole in the ground, may yield instructive information over longer periods of time. Care must be taken to include corrections for any rainfall and soil water drainage which may occur between readings.

CHAPTER 5

Ground temperature and heat conduction

The surface layers of the earth are composed of a vast variety of materials ranging from water, ice and snow through a range of minerals found in rock, clay and sand to soils with various proportions of organic matter, air spaces and moisture. In opaque, non-permeable solids, heat transfer is by conduction but in porous materials such as snow, convective and turbulent transfer of sensible heat as well as vapour can be of major importance, whilst in clay, sand and soil, heat can be transported by percolating water as well as by conduction. The large transparent ice masses found in the coastal regions of Antarctica, for example, exhibit the interesting phenomenon of simultaneous transfer by radiation penetration and conduction. The water in oceans and lakes introduces a number of complex possibilities through combinations of conductive, radiative and turbulent heat transfer, the latter depending on stability, wind and tidal action. In some circumstances, both on land and in water, the occurrence of freezing can introduce further complications because of latent heat effects.

In all these materials, the vertical transfer of heat at any depth can be expressed as a flux dependent both on the vertical temperature gradient at that point and on the effective thermal conductivity. The latter conceptual parameter is identical with the actual thermal conductivity only when conduction is the only heat transferring process.

The conducted flux of heat, G, is related to the temperature gradient $\partial T/\partial z$ by:

$$G = k \frac{\partial T}{\partial z} \qquad [5.1]$$

where k is the thermal conductivity. The value of this parameter for a number of terrestrial surface materials is shown in Table 5.1. In sand and soil for example, which may have substantial pore spaces, the conductivity depends greatly on whether the latter are filled with air or water but there is no simple functional relationship.

The heat capacity per unit volume, C, can be described more easily, being simply dependent on the relative proportions of constituents, rather than on their spatial arrangement as well, as in the case of thermal conductivity. Taking soil as an example, the effective specific heat per unit volume is given by:

TABLE 5.1

Thermal properties of natural materials

Substance	Thermal conductivity, k ($mW\,cm^{-1}\,{}^\circ C^{-1}$)	Volumetric specific heat, C ($J\,cm^{-3}\,{}^\circ C^{-1}$)	Thermal diffusivity, K ($cm^2\,sec^{-1}$)	Thermal admittance, $\sqrt{kC}$ ($W\,cm^{-2}\,{}^\circ C^{-1}\,sec^{1/2}$)
Quartz	90	1.2	0.075	0.33
Clay minerals	30	1.9	0.016	0.24
Organic matter	2.5	2.5	0.001	0.08
Water	5.8	4.2	0.0014	0.16
Ice	22	2.1	0.0105	0.68
Air	0.25	0.0013	~ 200	0.0006

Note: The effective thermal conductivity (and hence diffusivity) of water and air in motion can result in increases by orders of magnitude.

Natural media that are composed of any of the above listed materials incorporating air spaces have values of thermal conductivities significantly below that of the main solid matter. For such porous substances, the volumetric specific heat can be considered to be reduced in proportion to the density.

$$C = x_m C_m + x_o C_o + x_w C_w + x_a C_a \qquad [5.2]$$

where x_m, x_o, x_w and x_a are respectively the fractional proportions of minerals, organic matter, water and air in unit volume of soil and C_m, C_o, C_w and C_a are the corresponding specific heats which have been listed in Table 5.1. This table also shows the thermal diffusivity which is defined by:

$$K = k/C \qquad [5.3]$$

As will be seen later, the thermal diffusivity provides a measure of the response rate of a material to a change in temperature.

5.1. METHODS OF GROUND TEMPERATURE MEASUREMENT

All of the methods described in section 1.1 may find applicability. However, in measuring temperatures at various depths in the ground it is necessary to arrange either for remote indication or to grossly extend the thermal inertia of the sensors, allowing them to be withdrawn and read above ground level. The former approach is achieved by electrical temperature sensors being monitored by a suitable recorder, or by liquid thermometers, for example mercury-in-steel, hydraulically connected by flexible hoses to a recording or indicating mechanism. The latter technique of thermally lagging suitable indicating thermometers is exemplified by the often used mercury-in-glass thermometers with sensor bulbs encased in a substantial volume of wax whose high specific heat and low thermal conductivity combine to reduce the response time markedly. Such thermometers may be suspended by chain or cable at the desired depth in a borehole in the ground.

Under some circumstances, the presence of a borehole may constitute an undesirable perturbation. Generally, the matter of a rapid response time is unimportant in ground temperature measurements because the thermal mass of the surrounding medium damps out higher frequency fluctuations, although surface temperatures may constitute an exception.

The thermal coupling of a thermometer cannot be varied greatly in the ground, unlike the situation holding in the air or in a liquid, where forced ventilation or agitation increases the rate of heat flow between the sensor and the surrounding medium. This inability to increase the rate of dissipation of sensor energy is usually only serious when resistance thermometers are operated at too high a current level but is a problem with almost all types of sensors in transparent or translucent media such as ice and snow. Because solar radiation is able to penetrate these natural bodies to considerable depths, the temperature sensors used in such irradiated surroundings must be designed with some attempt at optical simulation. Dark, opaque thermometers used during daytime in ice or snow provide a more useful indication of radiation penetration than of temperature.

In general, ground temperature sensors should not interfere with the naturally occurring passage of light, air and moisture, yet be in good thermal contact with the surroundings.

5.2. THERMO-ELECTRIC EFFECTS

If a metal bar is heated at one end, the temperature there and hence free electron energy is increased to a level above that at the cooler end. The resultant diffusion of electrons causes an electro-motive force (e.m.f.), known as the Thomson e.m.f. Further, if two dissimilar metals are joined, electrons diffuse from one metal to another until a field of sufficient strength is established to maintain equilibrium. The junction is therefore the seat of an e.m.f. called the Peltier e.m.f. The magnitude of the Thomson e.m.f. is determined by the nature of the metal involved and the temperatures of the two points between which the e.m.f. is measured. The Peltier e.m.f. on the other hand depends on the nature of each of the two metals forming a junction and its temperature.

Referring to Fig. 5.1, it is seen that if a closed circuit is formed by joining the ends of two dissimilar metals A and B and holding the two paired extremities at temperatures T_1 and T_2, then the Thomson e.m.f.'s may be given by $\epsilon_T(T_1, T_2, A)$ and $\epsilon_T(T_1, T_2, B)$ and the Peltier e.m.f.'s by $\epsilon_P(T_1, A, B)$ and $\epsilon_P(T_2, A, B)$ where the nature of the functional dependence on temperatures and materials has been made clear. In this circuit, the two pairs of e.m.f.'s do not cancel unless $T_1 = T_2$. The resultant is known as the Seebeck e.m.f. which is given by:

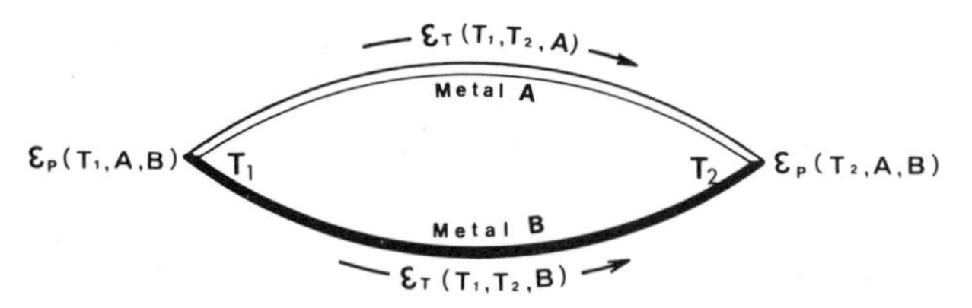

Fig. 5.1. The thermal electro-motive forces in a bi-metallic circuit.

$$\begin{aligned}\epsilon &= \epsilon_S(T_1, T_2, \text{A}, \text{B}) \\ &= \epsilon_P(T_1, \text{A}, \text{B}) + \epsilon_T(T_1, T_2, \text{A}) - \epsilon_P(T_2, \text{A}, \text{B}) - \epsilon_T(T_1, T_2, \text{B}) \qquad [5.4]\end{aligned}$$

This net e.m.f. can be measured by breaking the circuit at any point in either metal A or B. The network thus formed is known as a thermocouple and forms a simple reliable and important device for temperature measurement as it is not dependent on the physical dimensions of the conductors but only on their composition.

Accurate calibrations exist for many standard junctions of various pairs of metals and alloys, so that if one junction of a thermocouple is kept at the prescribed constant reference temperature, usually 0°C, the other may be used as a measuring probe. Table 5.2 lists the thermo-electric potential of a number of metals and alloys, allowing a selection to be made of the most suitable pair for specific purposes. Table 5.3 lists the Seebeck e.m.f.'s generated respectively by (a) copper-constantan and (b) iron-constantan thermocouple junctions.

TABLE 5.2

Thermo-electric potential of various metals, relative to lead, at 0°C

Substance	Thermo-electric potential ($\mu V\ ^\circ C^{-1}$)
Antimony	+ 35.6
Bismuth	− 43.7
Constantan (60% Cu, 40% Ni)	− 38.1
Copper	+ 1.3
Iron	+ 17.2
Nickel	− 23.3
Platinum	− 0.6
90% platinum and 10% rhodium	+ 6.4
Zinc	+ 3.0

5.3. THE THEORY OF GROUND HEAT CONDUCTION

The vertical temperature profile can readily be measured in the ground and at the surface but special devices and precautions are necessary for the direct monitoring of conducted heat flow. Hence in this section the relationship between ground temperature, T, and the associated heat fluxes, G, will be explored.

The fundamental equations of heat flow are Fourier's Law:

$$G = -kT' \qquad [5.5]$$

and the Principle of Heat Conservation:

$$G' = -C\dot{T} \qquad [5.6]$$

where k is the thermal conductivity and C is the specific heat per unit volume, or volumetric heat capacity. The differential notation employed is such that: the prime ($'$) denotes differentiation with respect to z, the depth axis and the dot ($\cdot$) denotes differentiation with respect to t, time.

From equation [5.5] differentiation yields:

$$G' = -kT'' \qquad [5.7]$$

so that on comparison with [5.6], it follows that:

$$\dot{T} = KT'' \qquad [5.8]$$

where $K = k/C$, the thermal diffusivity.

Further, from [5.5] and [5.6] it similarly follows that:

$$\dot{G} = -k\dot{T} \quad \text{and} \quad G'' = -C\dot{T}$$

Hence:

$$\dot{G} = KG'' \qquad [5.9]$$

Equations [5.8] and [5.9] show that identical mathematical expressions apply to both temperature and heat flow in a conductor and as a consequence it follows that similar solutions apply. In particular, if the temperatures can be described by a sinusoidal function, then a similar expression of the same frequency must apply to the heat flow, with the only possible differences being in amplitude and phase.

While the boundary conditions at the surface vary considerably, most natural circumstances can be described in terms of sinusoidal variations of surface temperatures, such as are exemplified by diurnal or annual events.

In the absence of advective or non-reversible thermal processes beneath the ground surface, it is important to note that for periodic processes of period τ, the mean flux of conducted heat at any depth is given by:

$$\bar{G} = \frac{1}{\tau}\int_{t}^{t+\tau} G\,dt = 0$$

TABLE 5.3

Thermocouple electro-motive forces (mV)*

T(°C)	0	1	2	3	4	5	6	7	8	9
(a) *Copper versus constantan thermocouple outputs*										
−50	−1.804	−1.838	−1.871	−1.905	−1.938	−1.971	−2.004	−2.037	−2.070	−2.103
−40	−1.463	−1.498	−1.532	−1.567	−1.601	−1.635	−1.669	−1.703	−1.737	−1.771
−30	−1.112	−1.148	−1.183	−1.218	−1.254	−1.289	−1.324	−1.359	−1.394	−1.429
−20	−0.751	−0.788	−0.824	−0.880	−0.897	−0.933	−0.969	−1.005	−1.041	−1.076
−10	−0.380	−0.417	−0.455	−0.492	−0.530	−0.567	−0.604	−0.641	−0.678	−0.714
(−) 0	0.000	−0.038	−0.077	−0.115	−0.153	−0.191	−0.229	−0.267	−0.305	−0.343
(+) 0	0.000	0.038	0.077	0.116	0.154	0.193	0.232	0.271	0.311	0.350
10	0.389	0.429	0.468	0.508	0.547	0.587	0.627	0.667	0.707	0.747
20	0.787	0.827	0.868	0.908	0.949	0.990	1.030	1.071	1.112	1.153
30	1.194	1.235	1.277	1.318	1.360	1.401	1.443	1.485	1.526	1.568
40	1.610	1.652	1.694	1.737	1.779	1.821	1.864	1.907	1.949	1.992
50	2.035	2.078	2.121	2.164	2.207	2.250	2.293	2.336	2.380	2.423
60	2.467	2.511	2.555	2.599	2.643	2.687	2.731	2.775	2.820	2.864
70	2.908	2.953	2.997	3.042	3.087	3.132	3.177	3.222	3.267	3.312
80	3.357	3.402	3.448	3.493	3.539	3.584	3.630	3.676	3.722	3.767
90	3.813	3.859	3.906	3.952	3.998	4.044	4.091	4.138	4.184	4.230

Table 5.3 (continued)

T(°C)	0	1	2	3	4	5	6	7	8	9
(b) *Iron versus constantan thermocouple outputs*										
−50	−2.43	−2.48	−2.52	−2.57	−2.62	−2.66	−2.71	−2.75	−2.80	−2.84
−40	−1.96	−2.01	−2.06	−2.10	−2.15	−2.20	−2.24	−2.29	−2.34	−2.38
−30	−1.48	−1.53	−1.58	−1.63	−1.67	−1.72	−1.77	−1.82	−1.87	−1.91
−20	−1.00	−1.04	−1.09	−1.14	−1.19	−1.24	−1.29	−1.34	−1.39	−1.43
−10	−0.50	−0.55	−0.60	−0.65	−0.70	−0.75	−0.80	−0.85	−0.90	−0.95
(−) 0	0.00	−0.05	−0.10	−0.15	−0.20	−0.25	−0.30	−0.35	−0.40	−0.45
(+) 0	0.00	0.05	0.10	0.15	0.20	0.25	0.30	0.35	0.40	0.45
10	0.50	0.56	0.61	0.66	0.71	0.76	0.81	0.86	0.91	0.97
20	1.02	1.07	1.12	1.17	1.22	1.28	1.33	1.38	1.43	1.48
30	1.54	1.59	1.64	1.69	1.74	1.80	1.85	1.90	1.95	2.00
40	2.05	2.11	2.16	2.22	2.27	2.32	2.37	2.42	2.48	2.53
50	2.58	2.64	2.69	2.74	2.80	2.85	2.90	2.96	3.01	3.06
60	3.11	3.17	3.22	3.27	3.33	3.38	3.43	3.49	3.54	3.60
70	3.65	3.70	3.76	3.81	3.86	3.92	3.97	4.02	4.08	4.13
80	4.19	4.24	4.29	4.35	4.40	4.46	4.51	4.56	4.62	4.67
90	4.73	4.78	4.83	4.89	4.94	5.00	5.05	5.10	5.16	5.21

* Reference junction 0°C.

On the other hand, the mean ground temperature at a given depth is:

$$\overline{T} = \frac{1}{\tau} \int_{t}^{t+\tau} T dt \neq 0 \text{ in general.}$$

If consideration is restricted to one frequency alone, the standard solution of equation [5.8] is for the following surface temperature boundary condition:

$$T_0(t) = \overline{T_0} + \Delta T_0 \cos(nt - \delta) \qquad [5.10]$$

[where $T_0(t) = T(0, t)$, i.e. $z = 0$] at any depth z below the surface:

$$T(z, t) = \overline{T_0} + \Delta T_0 e^{-z/Z} \cos\left(nt - \frac{z}{Z} - \delta\right) \qquad [5.11]$$

where $Z = \sqrt{2K/n}$ and n is the sinusoidal frequency. More general solutions can be built up by summing over the various frequencies that may be relevant, allowing for individual amplitudes and phase differences.

From [5.5], it follows that:

$$G(z, t) = -k \frac{\partial T(z, t)}{\partial z}$$

so that from [5.11]:

$$\frac{\partial T(z, t)}{\partial z} = -\Delta T_0 \left[\frac{1}{Z} e^{-z/Z} \cos\left(nt - \frac{z}{Z} - \delta\right) + \frac{1}{Z} e^{-z/Z} \sin\left(nt - \frac{z}{Z} - \delta\right)\right]$$

$$\therefore G(z, t) = \frac{k \Delta T_0 e^{-z/Z}}{Z} \left[\cos\left(nt - \frac{z}{Z} - \delta\right) + \sin\left(nt - \frac{z}{Z} - \delta\right)\right]$$

which expression can be further simplified to:

$$G(z, t) = \frac{\sqrt{2} k \Delta T_0 e^{-z/Z}}{Z} \cos\left(nt - \frac{z}{Z} - \delta - \frac{\pi}{4}\right) \qquad [5.12]$$

because $\cos \pi/4 = \sin \pi/4 = 1/\sqrt{2}$.

Simply by comparing equations [5.11] and [5.12] it can be seen that at any depth, the temperature and heat flux waves are out of phase by $\pi/4$.

If the amplitudes of the temperature and heat flux waves at depth z are respectively given by ΔT and ΔG, then from equations [5.11] and [5.12] it is readily seen that:

$$\Delta T/\Delta T_0 = \Delta G/\Delta G_0 = e^{-z/Z} \qquad [5.13]$$

Equation [5.13] implies that for a given frequency of a thermal wave in ground of given diffusivity, an exponential attenuation takes place with depth. Furthermore, recalling that $Z = \sqrt{2K/n}$, it is evident that the rate of attenuation is less for larger values of the diffusivity and greater for higher frequencies.

At the surface:

$$G_0(t) = G(0, t)$$

$$= \frac{\sqrt{2}}{Z} k \Delta T_0 \cos\left(nt - \delta - \frac{\pi}{4}\right)$$

$$= \sqrt{nkC} \Delta T_0 \cos\left(nt - \delta - \frac{\pi}{4}\right) \quad [5.14]$$

$$\therefore \Delta G_0 = \sqrt{nkC} \Delta T_0$$

$$\text{and } \Delta G_0 / \Delta T_0 = \sqrt{nkC} \quad [5.15]$$

From equation [5.15] it follows that for a given frequency, the ratio of amplitudes of surface heat flux and surface temperature waves is proportional to $\sqrt{kC}$ a term variously known as the thermal admittance or the conductive capacity. Mathematically, it is the geometric mean of the thermal conductivity and the volumetric heat capacity. Typical values are shown in the last column of Table 5.1.

Because ground temperature records are the resultant of several frequency components, it is important to enlist a suitable filtering technique to select single frequencies for analysis on the above lines.

The value of $Z = \sqrt{2K/n}$ can be found from equation [5.13], i.e. by the comparison of amplitudes of a single-frequency temperature wave at various depths. An alternative method involves comparison of the phases of this wave as a function of depth and referring to equation [5.11]. If the ground is thermally inhomogeneous, the two methods will show discrepancies.

The determination of Z for a given wave frequency n allows the calculation of the thermal diffusivity K from which latter term the conductivity k can be calculated if the volumetric heat capacity C can be measured or estimated. Measurement of the heat capacity of a sample of the ground medium involves calorimetry. It is necessary to know the thermal conductivity to be able to calculate the amplitudes of the heat flux wave from the temperature data.

Experiment XXV. Determination of thermal diffusivity from temperature profile observation

A probe of six thermo-junctions spaced at 5-cm intervals and soldered to small brass cylindrical sections continuously insulated and protected by a moisture-proof polythene tube should be constructed. A seventh junction

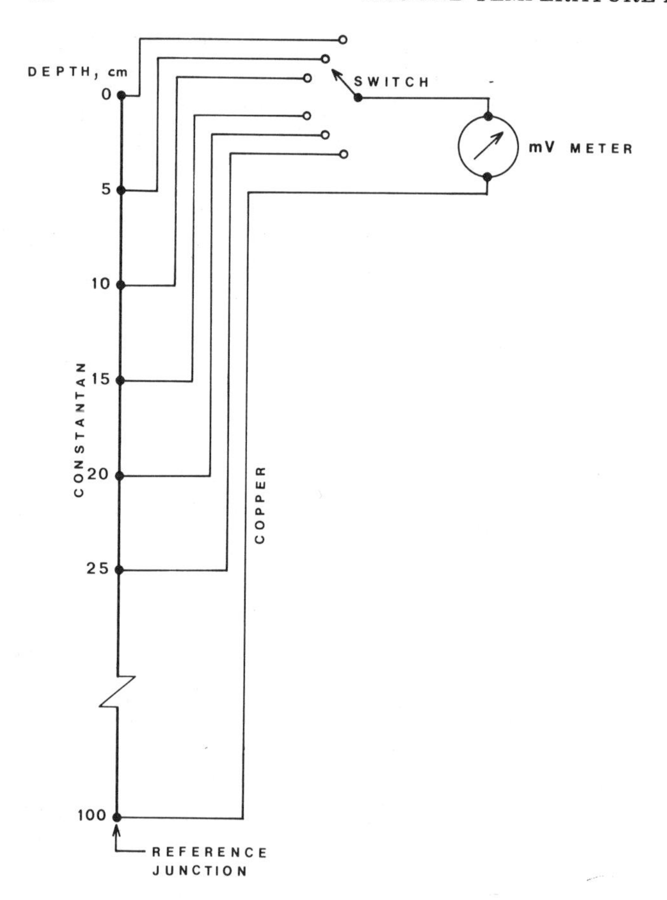

Fig. 5.2. Connections for a linear multi-junction thermocouple probe. A constantan wire runs the full length of the probe with junctions with copper formed by soldering at various points. Ideally, the switch and millivolt meter should be replaced by an automatic multi-channel millivolt recorder. At a depth of 100 cm, the lowest junction is deep enough to escape any diurnally induced temperature variation in soil. The junctions must be electrically, but not thermally, insulated from the ground. This can be accomplished by soldering the junctions to small (< 1 cm long) brass cylinders, through which all the conductors can be passed. These cylinders are then located within a well-fitting, thin-walled PVC tube running the length of the probe and sealed at the lower end.

should be connected to act as a reference for the other six. As shown in Fig. 5.2, the probe should be buried in the ground with its axis vertical so that the first six junctions are able to respond to soil temperatures at 0, 5, 10, 15, 20 and 25 cm depths with reference to the seventh at 1 m. The e.m.f.'s can be monitored either by manual switching and suitably frequent meter reading or preferably by a 6-channel chart recorder. Alternatively, two three-level hydro-mechanical thermographs can be used.

If $\overline{T}_1'$ and $\overline{T}_2'$ are the mean temperature gradients at two depths z_1 and

z_2 below the ground over a period of time Δt then the change in heat content of the layer between z_1 and z_2 is given by the difference in heat fluxes at the upper and lower levels. On the assumption that the thermal conductivity, k, and volumetric heat capacity C are uniform, it follows that:

$$k(\bar{T}_1' - \bar{T}_2') = C(z_2 - z_1)\frac{\Delta\bar{T}}{\Delta t} \qquad [5.16]$$

where $\Delta\bar{T}$ is the change in the mean temperature, $\bar{T}$, of the layer $\Delta z = z_2 - z_1$, under observation during the time t. Equation [5.16], when written as:

$$\frac{\Delta\bar{T}}{\Delta t} = K\frac{\Delta\bar{T}'}{\Delta z}$$

where $\Delta\bar{T}' = \bar{T}_1' - \bar{T}_2'$, may be recognized as a finite difference form of the second-order differential equation [5.8], the thermal diffusion equation.

It T_1', T_2', z_1, z_2, $\Delta\bar{T}$ and Δt are measured, then the thermal diffusivity can be calculated and is given by:

$$K = \frac{k}{C} = \frac{(z_2 - z_1)\Delta\bar{T}}{(\bar{T}_1' - \bar{T}_2')\Delta t} \qquad [5.17]$$

Using the thermocouple probe suggested, mean temperature gradients can be found at depths of 5 and 20 cm, for example over a period of about 2 hours. The weakness of this method lies in the difficulty of determining the appropriate values for the mean temperature gradients $\bar{T}_1'$ and $\bar{T}_2'$ with adequate precision.

Experiment XXVI. Diurnal temperature and heat flux waves in the ground

The same probe as described for use in Experiment XXV should again be employed. One advantage of the scheme in which temperatures at levels between 0 and 25 cm in depth are referred to a reference at 1 m is that longer period variations such as result in a gradual trend in soil temperatures over a week, for example, are compensated out. The record includes only those higher frequencies which do not penetrate significantly to the 1 m level.

Because the diurnal pattern of irradiational heating by the sun even under clear sky conditions is not purely sinusoidal, the temperature induced at the ground surface consists of several important contributing frequencies. Therefore, in order to attempt an analysis of the observed temperatures by equations such as [5.11], the main contributing frequencies must be separated.

The simplest method of selecting a working frequency sequence from the data is to take a 12-hour running mean. This process will result in the elimination of all constant amplitude components having periods $= 12/n$, where n is an integer $\geqslant 1$. This procedure can either be carried out with a suitable electronic calculator or by a simple computer program.

This experiment is ideally performed over a continuous sequence of clear days in late summer so that the various frequency components will remain nearly constant in amplitude and the mean diurnal soil temperature will not change significantly during the course of the measurements, although the latter can be compensated for instrumentally as noted above. As a result, there will be no important components with a period $\geqslant 24$ hours in the data. The numerical filtration technique described will then yield a practically pure record of 24-hour period.

5.4. HEAT FLUX METERS

Although ground heat fluxes may be measured by the method of Experiment XIX, this indirect approach is not always satisfactory. Non-isotropic media introduce special problems which can, however, be solved. The most serious limitation results from the need for several cycles of the main contributing frequencies to be observed to enable a reliable computation of the thermal diffusivity, a quantity which often varies under a range of commonly occurring transient conditions. Rapid change of thermal properties of ground media occur for example when soil moisture content changes either as a result of rainfall or evaporation.

Passive heat flow transducers in one form or another are used as sensors of both conducted and radiated heat. In the former application, the sensor is generally known as a heat flux meter (HFM) and in the latter, for example, as a net radiometer. There are also various types of active sensors, which depend on the generation of a perturbing or in some cases compensating source of heat. Such devices can also be employed with suitable modification either as HFM's or radiometers.

There are two ways of describing the thermal response of a passive sensor surrounded by a medium whose undisturbed temperature distribution in the absence of the sensor is known. One is to state the actual change in temperatures occurring in the region occupied by the sensor, the other is to compare the flow of heat through the sensor with that in the undisturbed surrounding medium. It is clear that this initial problem is greatly simplified if the undisturbed heat flow is one-dimensional, corresponding to the ideal conditions usually sought for other micro-meteorological flux measurements.

Passive thermal sensors of conductive phenomena can be classed between two extremes. One is exemplified by the exact HFM, a two-dimensional device oriented normally to the heat flow, the other is the exact temperature probe, a one-dimensional sensor monitoring temperature along its length.

Unfortunately, as will be illustrated by Experiment XXIX, real thermal

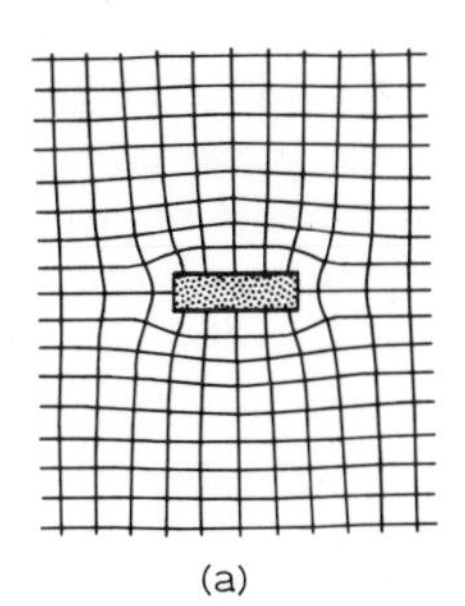

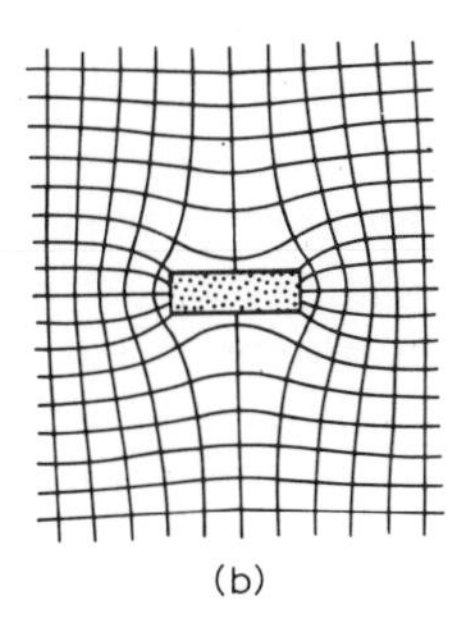

Fig. 5.3. Isotherms and heat flow lines in the vicinity of a heat flux meter thermally perturbing a conductive environment which elsewhere has a constant temperature gradient. These isotherms were determined by analogue measurements for a two-dimensional heat flux meter (uniform except for conducting covers) for the extreme cases of (a) infinite and (b) zero conductivity ratio.

sensors, in particular HFM's, only approximate to the extreme ideal. This fact is illustrated by Fig. 5.3 in which cross-sections of heat flow lines and isotherms are shown for a conducting material where the thermal conductivity of the meter is respectively greater and less than that of the surroundings. It is thus clear that good HFM design seeks to avoid the problem of what could be described as edge effects varying with relative conductivities. Because natural materials, including soils, can incorporate varying proportions of solids, air and water, a given HFM can be anticipated to experience relative changes in the conductivity of the surrounding material of more than an order of magnitude.

In practice, a useful HFM is thin compared to the linear dimensions of its cross-sectional area. Edge effects are kept to a minimum by choosing a high-conductivity material across which the temperature difference associated with the conducted flow of heat being monitored can be measured and in fact regarded as proportional to the heat flux.

5.5. THERMOPILES

A convenient form of heat flow detector, whether used in a radiometer or an HFM is a thermopile. Although the sensitivity, which is defined as the sensor output per unit incident (or undisturbed) heat flux, increases as the relative conductivity of the thermopile decreases, HFM's require some sacrifice of sensitivity in order to achieve greater validity of the calibration for a range of environmental thermal conductivities. Thus the spaces between the multiple sets of series connected thermocouples that constitute a thermopile, are usually filled with some dense electrically insulating material.

Thermopiles can be built either by a careful assembly of individual pairs of thermo-junctions or more easily and conveniently by selective electroplating of a spiral of a thermo-electrically dissimilar metal wire wound on to a suitable former. For example, if constantan wire is wound on to a two-dimensional strip former, then copper plated over one side of the major width of the strip for the entire length, then a series of constantan-"more copper than constantan" junctions is formed. When wound into a spiral and glued between anodized aluminium (and hence electrically, but not thermally, insulating) cover plates and protected around the edge, a typical thermopile element suitable for HFM use can be made.

Experiment XXVII. Calibration of a heat flux meter

One suitable method of calibrating an HFM is by the fundamental expediency of converting it temporarily into a net radiometer and is not without problems. An alternative simple method involves the use of the aluminium disks used in Experiment II. It is necessary for the surface shapes and areas of the HFM to be identical to those of the heavy disks.

The method assumes that good thermal contact can be established between the HFM and the upper surface of a heated disk, whose measured rate of temperature fall after pre-heating and placement in an insulating surround, as before, can easily be related to a heat flux, which if the insulation is perfect, must pass entirely through the HFM. This heat flux should be determined by the method of Experiment II, although ventilation is not necessary unless higher rates of heat flow are sought. The heat flow through the HFM will, of course, not be constant during an observational period of disk cooling, so that the sensor output should be monitored, preferably continuously on a suitable strip chart recorder for subsequent comparison with the calculated heat flow derived from the disk. The calibration value of the thermopile HFM under test should be expressed in the form of millivolt per milliwatt ($mV\ mW^{-1}$).

Experiment XXVIII. Comparison of temperature and heat flux waves

Micro-meteorological heat flux measurements usually seek a value for the conducted heat flux at the surface of the ground, which is a difficult location for the accurate operation of HFM's. Because of this it is useful to install several HFM's in a vertical array below the surface and to compare the observed heat flux waves with the temperature waves measured at similar locations with a temperature probe.

The consistency of the two sets of observations should be checked by means of determining the ratio $\Delta T/\Delta G$ for several depths as well as the actual value $e^{-z/Z}$. Equation [5.13] will yield the thermal diffusivity $K = nZ^2/2$ as a function of the depth z.

Calculation of heat fluxes from temperature gradients requires knowledge of the thermal conductivity. In principle this can be obtained from the diffusivity provided that the volumetric specific heat is known. Although the latter might be estimated after sample inspection and density determination (and in the case of soils, a water content determination) careful calorimetry constitutes the only reliable method.

CHAPTER 6

Electrical analogue modelling of thermal processes

The accurate observation of natural heat transfer phenomena involves great care, so that the prior exploration of these by means of electrical analogues which may easily be set up in a laboratory can be of considerable value. Analogues can be devised to give the observer control over scaling factors involving both linear dimensions and time, the latter being an important consideration in the investigation of transient phenomena.

A great variety of analogue networks can be built on the analogy existing respectively between temperature and potential difference, heat flux and electrical current, and thermal and electrical conductivity and respectively capacity.

6.1. STEADY STATE HEAT CONDUCTION

Although terrestrial temperature distributions do not, in general, remain constant, implying that steady state heat conduction rarely happens, the phenomenon does occur over limited periods of time and in some situations represents a useful approximation to the mean heat transfer over a full cycle of transient events. When steady state conditions are assumed, capacitance becomes unimportant and only the conductive process remains relevant.

The study of capacitanceless, purely conductive heat transfer is important in the analysis of conduction through non-uniform media including layered soils and situations where the ground is covered by a snow cover. The above two examples can be dealt with by one-dimensional analysis and the appropriate electrical analogue would consist of a linear array of electrical resistors. In this situation, Fourier's Law describing one-dimensional heat conduction:

$$G = -k\frac{\partial T}{\partial z} \qquad [6.1]$$

and Ohm's Law:

$$i = -\frac{1}{r}\frac{\partial V}{\partial x} \qquad [6.2]$$

are the analogous equations, with the heat flux G and electrical current i, the temperature T and potential difference V, the thermal conductivity k and the reciprocal resistivity per unit length, $1/r$, being respectively related

TABLE 6.1

Electrical simulation of thermal parameters involved in conduction

Parameter		Scaling factor	
thermal	electrical		
(a) *One-dimensional model*			
Temperature (°C)	Potential (V)	$f_{(TV)}$	V °C^{-1}
Heat flux (W cm^{-2})	Current (A)	$f_{(Gi)}$	A W^{-1} cm^{2}
Conductivity (W cm^{-1} °C^{-1})	Reciprocal resistance per unit length (Ω^{-1} cm)	$f_{(kr^{-1})}$	Ω^{-1} W^{-1} °C cm^{2}
Distance (cm)	Distance (cm)	$f_{(zx)}$	dimensionless
(b) *Two-dimensional model*			
Temperature (°C)	Potential (V)	$f_{(TV)}$	V °C^{-1}
Heat flux (W cm^{-2})	Current density (A cm^{-1})	$f_{(GJ)}$	A W^{-1} cm
Conductivity (W cm^{-1} °C^{-1})	Reciprocal resistivity (Ω^{-1})	$f_{(kR^{-1})}$	Ω^{-1} W^{-1} cm °C
Distance (cm)	Distance (cm)	$f_{(zx)}$	dimensionless

Note: In a three-dimensional model, the current density would be in A cm^{-2} and the resistivity in Ω cm.

parameters while the distances z and x are similar in both equations. The choice of electrical circuit parameters influences the scaling factors, including that for distances in the thermal and electrical processes. Table 6.1a summarizes the various relationships.

Referring to this table and equation [6.1] or [6.2], the sole constraint on the choice of scaling factors is seen to be given by:

$$f_{(Gi)} = \frac{f_{(kr^{-1})} f_{(TV)}}{f_{(zx)}} \qquad [6.3]$$

Since there is usually no good reason for not having $f_{(zx)} = 1$, it follows that $f_{(Gi)} = f_{(kr^{-1})} f_{(TV)}$. If heat conduction in a certain material were simulated by an electrical circuit in which the potential at any point in V were chosen to directly simulate temperature in °C and the heat flux in mW cm^{-2} by a directly analogous current in mA, then $f_{(TV)} = f_{(Gi)} = f_{(kr^{-1})} = 1$, so that if the conductivity of the thermal material were, for example, 20 mW cm^{-1} °C^{-1}, then the electrical circuit would require a resistance of 50 Ω cm^{-1}. On the other hand, if a reduced simulating voltage were desired, for example to ensure the safety of the observer, $f_{(TV)}$ could be chosen to be, say 0.1 V °C^{-1}. Thus if it were still required that $f_{(Gi)} = f_{(zx)} = 1$, then, $f_{(kr^{-1})} = 10$. In this case, the same conductivity of 20 mW cm^{-1} °C^{-1} would be simulated electrically by a resistance of 5 Ω cm^{-1}.

Two-dimensional heat conduction requires a two-dimensional electrical analogue for simulation. This can be provided by conductive (or resistive) paper, although the creation of conductive inhomogeneities requires some ingenuity. On the other hand, a matrix of simple resistors provides an electrical finite difference analogue. The latter can also be extended to examine three-dimensional problems while the continuous analogue approach needs to resort to conducting solutions. While many three-dimensional heat flow problems cannot be transformed to or cross-sectionally represented by a two-dimensional case, examination of the latter nevertheless can be either instructive, or in some cases even constitute a useful approximation. An example is given by the problem of a heat flux meter (HFM) which has been introduced in section 5.4. It is not unreasonable to construct an HFM in the form of a right-angled parallelipiped with relative dimensions similar to a knife blade. If such a blade is placed so as to intercept heat flow with its maximum cross-sectional area, then with the exception of the two far ends, two dimensions suffice for the description of the heat flow.

6.2. THE PERFORMANCE OF A HEAT FLUX METER

If an HFM is designed to monitor an undisturbed one-dimensional heat flow, the provision of thin, highly conductive cover plates over the meter's main cross-sectional surfaces allows the measurement of a representative temperature gradient across its thickness when installed in its surroundings (such as soil). If the temperature gradients through the HFM and in the undisturbed surroundings are denoted respectively by T'_m and T'_s and the heat fluxes and thermal conductivities are respectively G_m, G_s and k_m and k_s, then through the meter:

$$G_m = -k_m T'_m \qquad [6.4]$$

and through the undisturbed surroundings:

$$G_s = -k_s T'_s \qquad [6.5]$$

$$\therefore \frac{G_m}{G_s} = \eta \frac{T'_m}{T'_s} \qquad [6.6]$$

where $\eta = k_m / k_s$, the conductivity ratio.

Because the conductivities of natural materials vary not only between different substances but for specific media such as soil, whose conductivity can change by an order of magnitude depending on compaction and water content, it is essential to examine the ratio G_m / G_s as a function of η. Although this functional relationship is neither simple, nor unique, in depending on the geometry of the HFM and on its proximity to the surface of the surrounding medium, it is possible to illustrate the main tendencies by considering two extremes of conductivity ratio, assuming that an HFM is infinitely distant from all boundaries in the surroundings:

$$G_m / G_s \rightarrow a \text{ as } \eta \rightarrow \infty \qquad [6.7]$$

since the HFM can only extract a finite heat flux from the surroundings when the conductivity of the latter remains finite, and:

$$T'_m / T'_s \to b \text{ as } \eta \to 0 \qquad [6.8]$$

since a finite temperature difference remains across even a perfectly insulating object, where both a and b are constants depending on the geometry.

It can be shown that the simplest continuous expressions obeying the constraints of [6.6], [6.7] and [6.8] are:

$$\frac{T'_m}{T'_s} = \frac{1}{1 + H(\eta - 1)} \qquad [6.9]$$

and:

$$\frac{G_m}{G_s} = \frac{\eta}{1 + H(\eta - 1)} \qquad [6.10]$$

where $H = 1/a$ and $b = a/(a-1)$. The term H is called the geometric characteristic parameter of an HFM, thus named because it depends only on the shape of the meter. Equations [6.9] and [6.10] are clearly not independent and represent two alternatives for the same specification.

The parameter H can only assume values between 0 and 1. The case $H = 0$ represents the ideal temperature probe, since equation [6.9] shows $T'_m = T'_s$, i.e. the meter records precisely the same temperature distribution as occurs in the unperturbed surroundings. On the other hand, the ideal HFM is achieved when $H = 1$, for which equation [6.10] shows $G_m = G_s$, i.e. the value of the flux passing through the meter is the actual value of the undisturbed flux. However, there is no simple functional relation between an arbitrary, even regular geometrical shape of HFM and the corresponding geometric characteristic parameter. Fig. 6.1 illustrates the behaviour of HFM's for (a) temperature gradient and (b) flux specifications respectively.

Experiment XXIX. Analysis of the steady state response of a heat flux meter — using conducting paper

Highly conductive silver paint can be applied to the two opposite edges of a rectangular sheet of electrically conductive (or rather resistive) paper to ensure an undisturbed electrical field of uniform gradient when a potential difference is applied to the prepared edges. A two-dimensional HFM can be simulated electrically by initially silver painting in two parallel cover plates (in one-dimensional cross-section) using a tubular precision drawing pen. If the portion of paper sandwiched between the cover plate lines is now cut out, an HFM of zero thermal conductivity can be simulated. An alternative approach is to silver paint-in continuously the two-dimensional cross-sectional shape of an HFM, thus enabling the characteristics of an infinitely

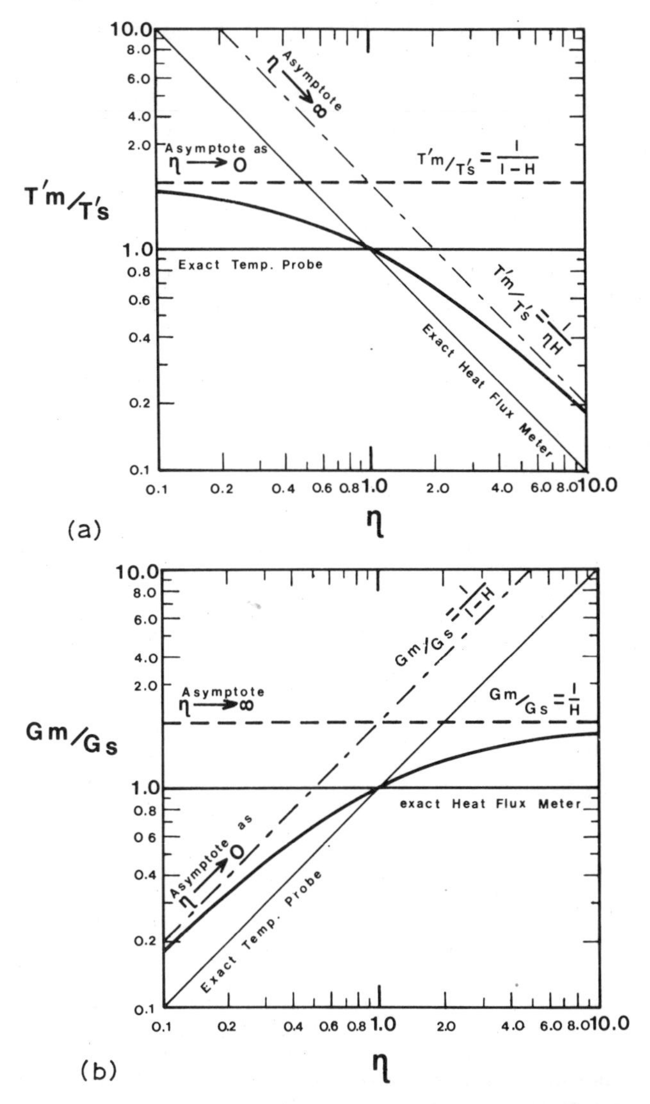

Fig. 6.1. (a) The dependence on conductivity ratio η of the sensitivity to temperature gradient of a heat flux meter. T_m' and T_s' are the temperature gradients across the meter and the undisturbed surroundings respectively. The specific example shown is for the geometric parameter $H = 0.5$. (b) The dependence on conductivity ratio η of the sensitivity to heat flux of a heat flux meter. G_m and G_s are the heat fluxes through the meter and the undisturbed surroundings respectively. The specific example shown is for the geometric parameter $H = 0.5$.

conducting HFM to be determined and then to transfer to the other extreme by cutting out all except the cover plate lines. The cutting is best done with a sharp knife. Intermediate values can now be formed by the connection of known values of electrical resistance.

A simple potentiometer, with a conducting lead pencil probe, suffices for drawing in isopotential lines (simulating isotherms) and the measurement of gradients. The arrangement is shown in Fig. 6.2.

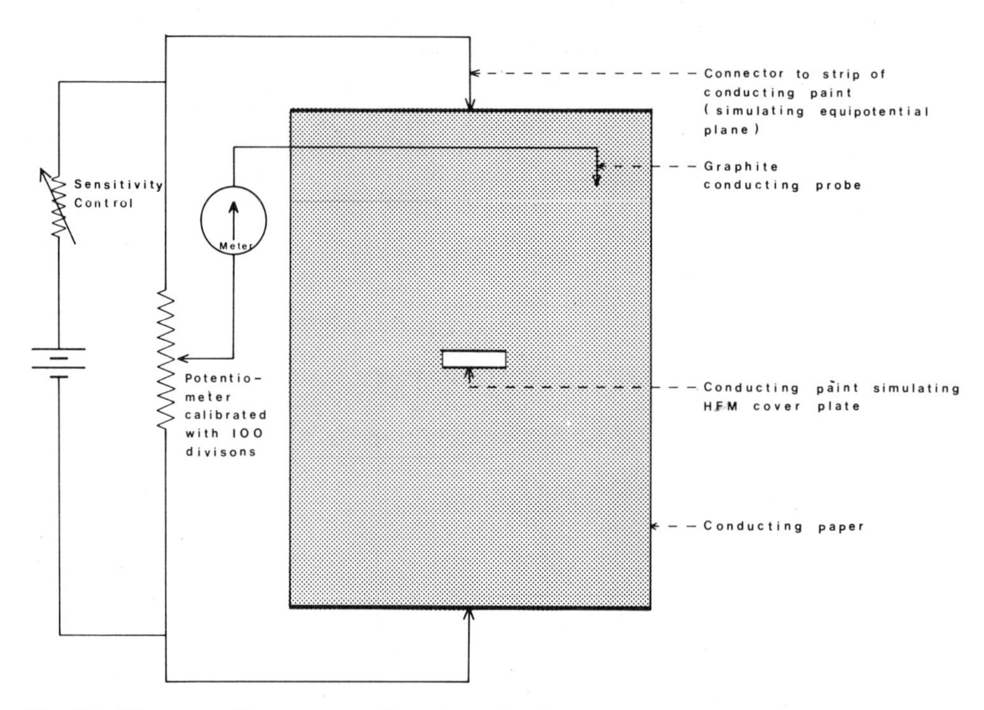

Fig. 6.2. Diagrammatic representation of conducting paper analogue for heat flux meter analysis. A uniform electric field is set up over a conducting sheet by connecting an electrical potential to two parallel opposite edges along which a thin strip of conducting paint has been applied. After testing this prepared field for uniformity, an area representing the cross-section of the heat flux meter under study is cut out and conducting paint is applied to simulate the cover plates. A variable resistor connected between these latter conducting strips allows controlled variation of the ratio of the effective conductivity of the heat flux meter area to that of the surroundings, represented by the conducting paper. Isopotential lines on the paper are located by moving the conducting probe in a locus over the paper so that a null meter reading is maintained for a given potentiometer setting.

Since only the relative conductivity η of the HFM to its surroundings need be specified, the actual conductivity of the conducting paper is irrelevant although it should be constant if a number of separate determinations are to be compared.

The described measurements of potentials V_m' and V_s', simulating T_m' and T_s' allow the preparation of a graph in the form of Fig. 6.1a for a given selected HFM shape and also the calculation of the parameter H. If several experimenters in a group are able to find the values of H for different rectangular shapes, pooling of the results would allow a graphical representation of the functional relationship between H and w/d, where w and d are the width and thickness of the HFM.

Although theoretically, the model HFM should be constructed at an infinite distance from surrounding boundaries, in actual practice a distance large compared to the dimensions of the meter suffices.

A valuable extension to this experiment involves examining the response of HFM of given shapes as a function of proximity to the upper boundary (surface), because it is there that HFM's are most frequently employed. At the surface, the upper face of an HFM is no longer in contact with what may be a thermally dissimilar material, giving rise to a theoretically relatively complex situation.

6.3. THERMAL DIFFUSION

The existence of a finite thermal capacitance means that even an infinite rate of change of temperature (e.g. a step function) at the boundary of a conductor, results in only a finite rate of change of internal temperatures. As has already been seen, the appropriate equation for one-dimensional heat flow in a homogeneous conductor is (equation [5.8]):

$$\dot{T} = \frac{k}{C} T''$$

where $K = k/C$, the thermal diffusivity and k is thermal conductivity and C the volumetric heat capacity. The behaviour of a one-dimensional electrical analogue is given by:

$$\dot{V} = \frac{1}{r\gamma} V'' \qquad [6.11]$$

where r and γ are respectively the electrical resistance and capacitance per unit length of conductor and as before, ($\dot{}$) and ($''$) imply $\partial/\partial t$, and $\partial/\partial x^2$.

It is clear that the electrical analogue is simply an extension of the simple conductor described in section 6.1 except that continuously distributed capacitance is now required. Because of the difficulty of achieving the latter for sufficiently large values of γ, one alternative is to approximate by means of finite values of capacitance at fixed intervals along the length of the conductor, resulting in the type of circuit illustrated in Fig. 6.2.

As before, various scalings are possible and summarized in Table 6.2. Some of the scaling factors are dimensionless but their magnitude can be chosen to suit a particular investigation.

From equations [5.8] and [6.11] it follows that:

TABLE 6.2

Electrical analogues of thermal parameters involved in diffusion (one-dimensional)

Parameter		Scaling factor	
thermal	electrical		
Temperature (°C)	Potential (V)	$f_{(TV)}$	V °C^{-1}
Heat flux (W cm^{-2})	Current (A)	$f_{(Gi)}$	A W^{-1} cm^{2}
Conductivity (W cm^{-1} °C^{-1})	Reciprocal resistance per unit length (Ω^{-1} cm)	$f_{(kr^{-1})}$	Ω^{-1} W^{-1} cm^{2} °C
Capacity (J °C^{-1} cm^{-3})	Capacitance per unit length (F cm^{-1})	$f_{(c\gamma)}$	F J^{-1} cm^{2} °C
Diffusivity (cm^2 sec^{-1})	Diffusivity (cm^2 sec^{-1})	$f_{(KD)}$	dimensionless
Distance (cm)	Distance (cm)	$f_{(zx)}$	dimensionless
Time (sec)	Time (sec)	$f_{(tT)}$	dimensionless

$$f_{(KD)} = 1/f_{(kr)}\, f_{(c\gamma)} \qquad [6.12]$$

and:

$$1/f_{(tT)} = f_{(KD)}/f^2_{(zx)} \qquad [6.13]$$

showing that if $f_{(zx)} = 1$, then the time scaling factor in the analogue can be selected by the appropriate choice of resistive and capacitive components. Equation [6.3] also still applies.

As an example of time scaling, if diurnal heat conduction phenomena of a period of 8.64×10^4 seconds are to be simulated electrically, an analogue period of 8.64×10^{-3} might be regarded as desirable, primarily because the latter can easily be displayed on an oscilloscope.

This involves a value of $f_{(tT)} = 10^{-7}$ and hence, if $f_{(zx)} = 1$, $f_{(KD)} = 10^7$. Thus if the thermal diffusivity of the medium being simulated is 10^{-2} cm sec^{-1}, the electrical analogue would require a diffusivity of 10^5 cm^2 sec^{-1}. Following the example cited in section 6.1, if it is chosen that $f_{(TV)} = 1$ and the conductivity of the thermal material is 20 mW cm^{-1} °C^{-1}, then of necessity the simulating resistances selected must be such that $r = 50\ \Omega$ cm^{-1}. However, since the electrical diffusivity is given by $1/r\gamma = 10^5$, the requirement for the electrical capacitance is that $\gamma = 1/50 \times 10^{-5}$ or 0.2 μF cm^{-1}.

Experiment XXX. Modelling of temperature waves in the ground

A one-dimensional diffusive electrical analogue network should be constructed for some terrestrial surface material. As an example, if values of the thermal conductivity and volumetric heat capacity have been obtained during Experiments XXV and XXVI, then these could be usefully used as a basis for the model. If diurnal temperature waves are to be examined, then a time scaling factor $f_{(tT)} = 10^{-7}$ would be appropriate.

The analogue circuit should be assembled by soldering the components to pins inserted in a strip of electronic matrix board. The number of finite difference elements in the circuit should be large enough to allow for simulation to a depth of 2 m which is well below the effective penetration of a diurnal wave. To a depth of 20 cm, 1-cm steps would be desirable, 10-cm increments should then extend to a depth of 1 m beyond which 50-cm differences suffice. The diagram is incorporated in Fig. 6.5.

The surface temperature wave should be provided by a suitable signal generator, set to the appropriate frequency and amplitude. The simulated temperatures should be read on an oscilloscope, which if available as a dual beam instrument can be used to make phase comparisons.

This apparatus suffices for the examination of both the attenuation and progressive phase shift of waves with depth, allowing the data obtained to verify the theoretical relationship of these parameters on the diffusivity (and frequency).

Although a conventional electronic signal generator is not able to provide a signal of the same wave form as a typical diurnal ground surface temperature record, an electrical sine wave can be suitable modified by the series connection of a diode by-passed with a resistor, the value of which determines the degree of asymmetry in the resultant wave. By the appropriate selection of components, a fair approximation to the shape of the diurnal wave can be achieved. The relatively rapid attenuation of the higher-frequency components of the asymmetrical wave can readily be observed in the analogue.

6.4. SIMULATION OF LATENT HEAT PROCESSES

In the simplest situations, a latent heat process such as evaporation might be considered to require only the provision of energy at a certain location or level, such as for example at the earth's surface. Indeed, to a first approximation no other physical change occurs, although in actual practice the thermal conductivity and volumetric heat capacity of the ground supplying the moisture for evaporation must change. In an electrical analogue of ground heat conduction for example, simulation of evaporation at the surface (and sensible heat transfer to the air) may be done by providing separate parallel circuits of a resistance adjusted so that the by-passed current is the analogue of the thermal fluxes.

The case of freezing of an ice layer floating on water represents a more complex modelling situation in that the thickness of the layer of ice, through which heat conduction must take place, changes as more water freezes at the lower interface. Thus the latent heat process actually changes the physical dimensions of the material transferring the heat involved and is illustrated by Fig. 6.3.

The analytical description of the above process is difficult and it is probable that an exact solution to the question of the rate of growth of ice as a function of time, temperature and thermal properties can be obtained only for the case in which the specific heat of the ice is assumed to be negligible. While this is certainly not the case, the assumption is not too unreasonable provided that ice temperatures are not far removed from the freezing point under which conditions the change in capacitive heat storage with new ice growth is negligible compared to the latent heat requirements for the freezing. The analytical formulation is often called "The Stefan Problem".

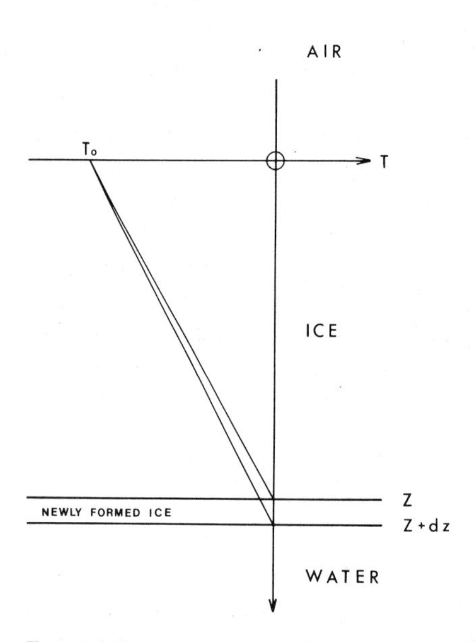

Fig. 6.3 Temperature gradients in growing, floating ice when heat capacity is neglected.

Fig. 6.3 indicates that with all temperature gradients linear, because with zero capacity the diffusivity is infinite, the temperature gradient in the ice at any time is given by:

$$T' = T_0/z$$

so that if the conductivity of the ice is k, the conducted flux of heat is:

$$G = -k\frac{T_0}{z} \qquad [6.14]$$

On the other hand, if the ice grows at the rate of dz/dt, then the flux of heat produced by the latent heat change must be conducted upward through the ice if the water below is uniformly at the freezing point so that:

$$G = -L\frac{dz}{dt} \qquad [6.15]$$

where L is the latent heat per unit volume of ice. Equations [6.14] and [6.15] imply that:

$$dz/dt = kT_0/Lz \qquad [6.16]$$

$$\therefore \int_0^h z\,dz = \frac{k}{L}\int_0^\tau T_0\,dt$$

$$\text{hence } h^2 = \frac{2k}{L}\int_0^\tau T_0\,dt \qquad [6.17]$$

where $z = h$ when $t = \tau$.

The integral on the right-hand side of equation [6.17] is conventionally referred to as the freezing exposure. The simplest case is for constant surface temperature T_0, when $h = \sqrt{2(k/L)\,T_0\tau}$, indicating a simple dependence of ice thickness, the rate of accretion of which falls off with the time elapsed.

A layer of "capacitanceless" floating ice can be simulated by a chain of resistors, the length of which must be added to at a rate proportional to the current flowing through the chain. Ideally this would be done by an automatic circuit which would monitor the current flow through the simulated ice-water interface. If each resistive increment represents, for example, 1 cm of ice, then when the integrated current flow totalled the simulated latent heat associated with 1 cm, then a new circuit element would be added. As shown in Fig. 6.4 the analogue method lends itself to the analysis of the more realistic case of ice having finite capacitance.

Experiment XXXI. The growth of ice floating on water

Ice formation is an important phenomenon and substantially influences the meteorology of polar regions where large areas of the oceans freeze annually. In making a simple electrical analogue of the process, it is preferable to initially model capacitanceless ice so that the experimental result can be compared with Stefan's exact analytical solution. Thus if the accuracy is acceptable in this simple case, capacitance can be added for a greater degree of realism.

Although the automatic simulation of freezing, based on current integration shown in Fig. 6.4 is ideal for the purposes of this experiment, such facilities may not always be available. However, if a time scaling factor of say $f_{(tT)} = 10^{-4}$ is adopted, addition of ice growth elements can be done by manually switching at times prescribed by the results of quick graphical integration of current recorded on a chart.

Because of the impossibility of measuring a finite gradient across a

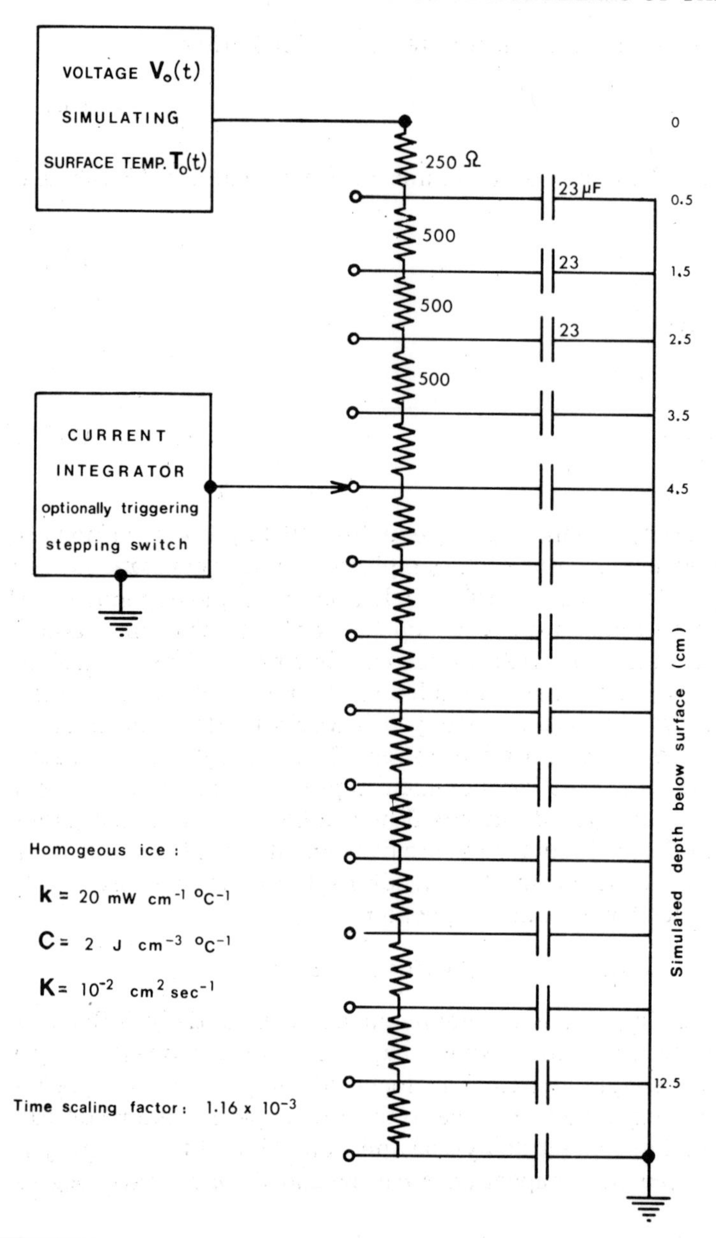

Fig. 6.4. An electrical analogue circuit for simulating the growth of floating ice. A current integrator may be connected to control the rate of addition of circuit elements automatically. Alternatively, the process may be performed manually. The values of circuit elements shown have been calculated to enable the latter and are based on the following scaling factors: $f_{(tT)} = 1/8.64 \times 10^3$, $f_{(Gi)} = 10$, $f_{(TV)} = 1$ and since $f_{(KD)} = 8.64 \times 10^3$ it follows that $f_{(kr^{-1})} = 10$. Hence if $k = 20$ mW cm^{-1} °C^{-1} and $C = 2$ J cm^{-3}, then $r = 500\ \Omega$ cm^{-1} and $\gamma = 23$ F cm^{-1}.

conductor of zero thickness, the model should commence with a simulated ice cover of 5 cm. For ice having a thermal conductivity of say 20 mW cm^{-1} $°C^{-1}$ and a volumetric latent heat of freezing of 300 J cm^{-3} (based on a latent heat of 333 J g^{-1} and a density of 0.9 g cm^{-3}), equation [6.16] shows that the rate of growth when the thickness is 5 cm is approximately 1.5×10^{-5} cm sec^{-1} $°C^{-1}$ which means that if the ice thickness is already 5 cm, then the growth rate for a surface temperature of 5°C below freezing point would be 1.3 cm day^{-1}. Hence if, in the electrical analogue, a time scaling factor $f_{(tT)} = 1/8.64 \times 10^3$ were introduced, each actual day would be simulated by a modelling time of 10 seconds. Thus if the analogue allows for 1-cm increments, the shortest time interval involved in an integration and resistive chain switching decision would be about 7.7 seconds. With two operators, one of whom should perform the continuing graphical integration as the current is recorded, the second should have no difficulty in switching at times sufficiently precise to introduce less than 5% error.

Care must be taken in the selection of values for the scaling factors $f_{(TV)}$ and $f_{(Gi)}$ to ensure realistic magnitudes for the capacitors in the circuit chain shown in Fig. 6.4.

In the analysis of freezing processes all temperatures are conveniently normalized so that the freezing point becomes 0° on a scale with Celsius intervals. This means that no radical circuit alterations are required when an analogue is adapted from fresh water freezing at 0°C to sea water freezing at −1.9°C.

6.5. SENSIBLE HEAT TRANSFER IN THE ATMOSPHERIC BOUNDARY LAYER

A similar analogue approach to that developed for the investigation of heat flow through solid ground can be employed to model heat transfer in the atmosphere, although the effects of turbulence and the convective properties of gases require special consideration. For example the upward transfer of heat into the atmosphere from a relatively warmer ground surface is a more efficient process than the corresponding case occurring when the sign of the temperature gradient is reversed. An unstable atmosphere transfers heat to a far greater extent than its stable counterpart.

Experiment XX has already demonstrated that the vertical gradient of the mean horizontal wind speed, $\partial \bar{u}/\partial z$, decreases with height and infers, through the association of equations [4.21], [4.22] and [4.23], that the same effect applies to temperature, as well as humidity, gradients. This is the basis of the further complication arising in the modelling of sensible heat transfer in the atmosphere in which the effective eddy conductivity of the air increases with height. The simplest approach is to assume that the turbulent transfer coefficient for sensible heat transfer, K_H, which is actually an effective diffusivity, can be related to height z in the same manner as the

TABLE 6.3

Dependence of the heat transfer coefficient (or eddy diffusivity) and the "effective eddy conductivity", $\rho c_p \bar{K}_H$, on the roughness parameter, z_0, and friction velocity, u_*. It has been assumed that the density of air, $\rho = 1.25 \times 10^{-3}$ g cm^{-3} and the specific heat at constant pressure, $c_p = 1.0$ J g^{-1} °C^{-1}

Location of layers (between z values)	$\bar{z}$ (geometric mean) [cm]	Δz [cm]	$\bar{K}_H$ ($= k u_* \bar{z}$) [cm^2 sec^{-1}]	$\rho c_p \bar{K}_H$ ($\rho c_p = 1.25 \times 10^{-3}$) [W cm^{-1} °C^{-1}]	$\rho c_p \bar{K}_H / \Delta z$ [W cm^{-2} °C^{-1}]	$\rho c_p \bar{K}_H (z_0, u_*)$	
						$z_0 = 0.1$ cm (smooth sand, soil) $u_* = 10, 100$ cm sec^{-1}	$z_0 = 10$ cm (short grass) $u_* = 10, 100$ cm sec^{-1}
$0, z_0$	(uniform laminar layers)	z_0	$0.4 u_* z_0$	$5 \times 10^{-4} u_* z_0$	$5 \times 10^{-4} u_*$	5×10^{-4}, 5×10^{-3}	5×10^{-2}, 5×10^{-1}
$z_0, 4z_0$	$2z_0$	$3z_0$	$0.8 u_* z_0$	$10^{-3} u_* z_0$	$\frac{1}{3} \times 10^{-3} u_*$	1×10^{-3}, 1×10^{-2}	1×10^{-1}, 1
$4z_0, 16z_0$	$8z_0$	$12z_0$	$3.2 u_* z_0$	$4 \times 10^{-3} u_* z_0$	$\frac{1}{3} \times 10^{-3} u_*$	4×10^{-3}, 4×10^{-2}	4×10^{-1}, 4
$16z_0, 64z_0$	$32z_0$	$48z_0$	$12.8 u_* z_0$	$1.6 \times 10^{-2} u_* z_0$	$\frac{1}{3} \times 10^{-3} u_*$	1.6×10^{-2}, 1.6×10^{-1}	1.6, 16

TABLE 6.4

The heat capacity of a volume Δz (cross-section of unit area through layer of thickness Δz) of air, where the layer thicknesses have been determined in Table 6.3

Location of layers (between z values)	$\bar{z}$ (geometric mean) [cm]	Δz [cm]	$\rho c_p \Delta z$ (for layer shown) $\rho c_p = 1.25 \times 10^{-3}$ J cm^{-3} °C^{-1}	$\overline{\rho c_p \Delta z}$ (geometric mean for adjacent layers)	$\overline{\rho c_p \Delta z}$ (J °C^{-1})	
					$z_0 = 0.1$ cm	$z_0 = 10$ cm
				0	0	0
$0, z_0$	(uniform laminar layer)	z_0	$1.25 \times 10^{-3} z_0$			
				$2.16 \times 10^{-3} z_0$	2.16×10^{-4}	2.16×10^{-2}
$z_0, 4z_0$	$2z_0$	$3z_0$	$3.75 \times 10^{-3} z_0$			
				$7.5 \times 10^{-3} z_0$	7.5×10^{-4}	7.5×10^{-2}
$4z_0, 16z_0$	$8z_0$	$12z_0$	$1.5 \times 10^{-2} z_0$			
				$3.0 \times 10^{-2} z_0$	3.0×10^{-3}	3.0×10^{-1}
$16z_0, 64z_0$	$32z_0$	$48z_0$	$6.0 \times 10^{-2} z_0$			
				$1.2 \times 10^{-1} z_0$	1.2×10^{-2}	1.2
$64z_0, 256z_0$	$128z_0$	$192z_0$	$2.4 \times 10^{-1} z_0$			

transfer coefficient for horizontal momentum, K_M. The latter is given by equation [4.25], so that in this elementary approach it is considered that:

$$K_H = u_* kz \qquad [6.18]$$

for heights $z \geqslant z_0$.

Equation [6.18] indicates that K_H, in addition to being height dependent, also increases with u_*, the friction velocity, which has already been shown to be dependent on the level of turbulence. As seen from equation [4.11], u_* increases with $\bar{u}$, for a fixed value of z. On the other hand, if z_0, which is a property of the ground surface, increases, then for a given $\bar{u}$ and z, u_* decreases.

It follows that a unique model of the atmosphere as a conductor of heat is impossible. Nevertheless, various situations for conditions of known u_* and z_0 can be examined.

In devising an analogue, the simplest method is to calculate the heights and thicknesses of atmospheric layers having equal thermal "resistances", a process which is illustrated by Table 6.3. For example, if $z_0 = 10$ and $u_* = 10$, it is seen that the mean "effective eddy conductivity" of the 30-cm air layer between 10 and 40 cm above the ground is 1 W cm^{-1} °C^{-1}. If the scaling factors employed in the second example of section 6.1 are applied, i.e. $f_{(Gi)} = f_{(zx)} = 1$, $f_{(TV)} = 0.1$ and hence $f_{(kr^{-1})} = 10$, it follows that eddy conductivity in the air layer under consideration would be simulated by an electrical resistance of 10 Ω^{-1} cm or 0.1 Ω cm^{-1}. The complete 30-cm layer would require 3 Ω in all. From Table 6.3 it can be seen that progressively thicker layers of 120, 480, . . . respectively above the 30-cm layer would also be simulated conductively by the same electrical resistance.

The complete analogue requires provision for simulating heat capacity and reference is made to Table 6.4 in which the heat capacity has been calculated for the layers corresponding to those devised in Table 6.3. Since single resistors are to be preferred for the individual layers, appropriate mean values of individual layer heat capacity have been calculated to enable simulating capacities of correct values to be connected at the terminals of resistors, since these would otherwise require an intermediate tapping if capacitive values were chosen to simulate directly those layers represented by the resistors.

Extending the earlier example for which $f_{(kr^{-1})} = 10$ and assuming (as in the example shown in section 6.3) that $f_{(tT)} = 10^{-7}$ so that $f_{(KD)} = 10^7$ (for $f_{(zx)} = 1$), it follows from equation [6.12] that $f_{(c\gamma)} = 10^{-8}$. This multiplier implies that a heat capacity of 7.5×10^{-3} J °C^{-1} is simulated by an electrical heat capacity of 7.5×10^{-11} F.

The circuit diagram for the above example is included in Fig. 6.5.

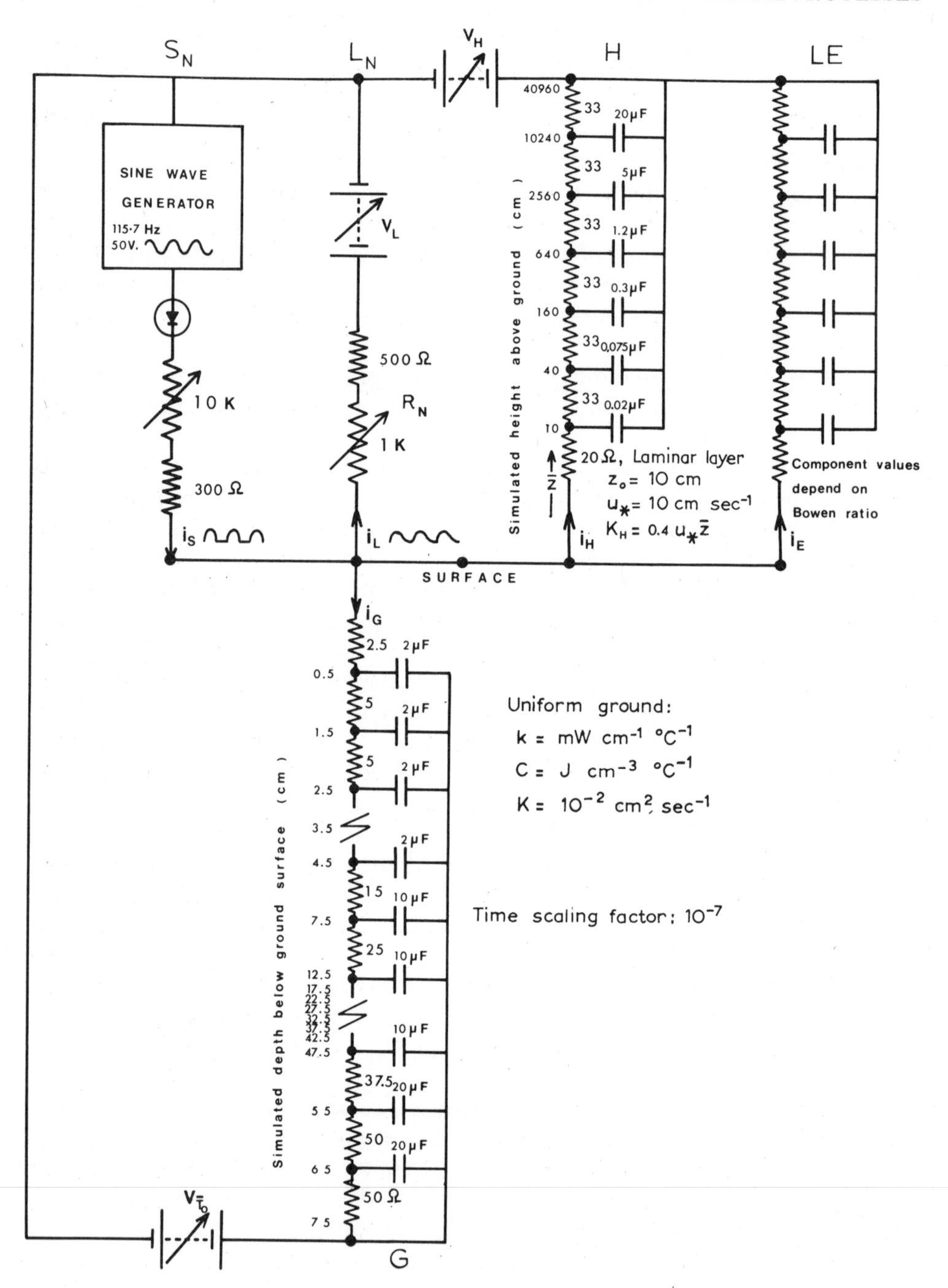

Fig. 6.5. An electrical micro-meteorological analogue with circuits for the modelling of net short-wave radiation (S_N), net long-wave radiation (L_N), ground heat conduction (G), atmospheric sensible heating (H) and evaporative equivalent heat flow (LE). The currents

6.6. LONG-WAVE RADIATION TRANSFER SIMULATION

The fluxes of radiation at the surface of the earth have been discussed in sections 2.3 and 2.4. Most terrestrial surfaces have long-wave emissivities of between 0.95 and 0.98, justifying an approximation of unity for this term in the application of the Stefan-Boltzmann Law (equation [1.9]) expressing the outgoing long-wave radiation L_o from the earth's surface at a temperature T_0:

$$L_o = \sigma T_0^4 \quad [6.19]$$

In seeking an electrical analogy for radiant transfer on the same lines as for ground heat conduction and atmospheric sensible heat transfer, a steady state solution is relatively easily derived. In this case, with L_0 and T_0 both constant, equation [6.19] can be written as:

$$L_o = \sigma T_0^3 \cdot T_0$$

$$\text{or } T_0 = \frac{1}{\sigma T_0^3} \cdot L_o \quad [6.20]$$

so that the term $1/\sigma T_0^3$ can be regarded as an "effective" radiative resistance" to a radiant flux L_o responding to a temperature difference T_0, the temperature being in °K.

Continuing to use analogue scaling factor values of $f_{(Gi)} = 1$ and $f_{(TV)} = 0.1$ as in the earlier examples in this chapter, an appropriate electrical analogue value for Stefan's constant, σ_e, follows as:

$$\sigma_e = f_{(Gi)} \cdot f_{(TV)}^{-4} \cdot \sigma = 10^4 \times \sigma \quad [6.21]$$

so that for $\sigma = 5.67 \times 10^{-2}\ \text{W cm}^{-2}\,{}^{\circ}\text{K}^{-4}$, $\sigma_e = 5.67 \times 10^{-8}\ \text{A V}^{-4}$. Thus in the electrical model of the radiative process, the "effective radiative resistance" $1/\sigma T_0^3$ is simulated by a resistor having the value:

$$R_0 = \frac{1}{\sigma \cdot T_0^3 \cdot f_{(Gi)} \cdot f_{(TV)}^{-1}} = \frac{10^{11}}{5.67 T_0^3} \quad [6.22]$$

Should, for example, $T_0 = 300$°K, then $R_0 = 650\ \Omega$.

Unfortunately, use of this type of analogue simulation will give rise to errors since the "effective radiative resistance" decreases with a rise in

---

i_S, i_L, i_G, i_H, and i_E respectively simulate the surface values of the fluxes S_N, L_N, G, H and LE. Typical wave forms for i_S and i_L have been shown. The potential $V_{\bar{T}_0}$ is set to simulate the mean ground temperature $\bar{T}_0$, similarly the value of V_H depends on the mean temperature of the air at the height to which the atmospheric heat flux model extends and V_L relates to the radiatively significant temperature T_i. Values of the resistive and capacitive components in the five flux simulating circuits must be calculated for specific conditions. The approximate relative wave shapes of the currents at various points in the circuit have been shown diagrammatically.

surface temperature. Thus, if T_0, the surface temperature, varies significantly about some mean value $\bar{T}_0$, chosen to select a fixed $\bar{R}_0$, then this value will be too large when $T_0 > \bar{T}_0$ and too small when $T_0 < \bar{T}_0$ resulting in the current flow simulating radiant flux being too small and too large respectively. The magnitude of this error may be reduced by selecting an appropriate thermistor-resistor combination whose total resistance will decrease as the current passed rises.

The above problem may be obviated by means of an electronic circuit, devised so as to pass a current proportional to the fourth power of the applied potential.

Experiment XXXII. A micro-meteorological model

A simple micro-meteorological regime can now be simulated for heat conduction, atmospheric sensible heating and long-wave radiation using an assembly of the analogue circuits that have been described in sections 6.3, 6.5 and 6.6 respectively.

One form of the heat balance equation is shown in equation [4.28]. If the net radiation term in this expression is replaced by:

$$A_N = S_N - L_N \qquad [6.23]$$

where S_N and L_N are the net short- and long-wave fluxes, with, it should be noted, opposite sign conventions. On substituting [6.23] equation [4.25] can be rewritten as:

$$S_N = L_N + G + H + LE \qquad [6.24]$$

In this new expression above, net long-wave radiation, ground heat conduction, sensible atmospheric heating and the vaporization flux can be interpreted as responding to the net short-wave radiation, a term which is completely determined by the incoming short-wave radiation and the albedos.

The right-hand side terms of equation [6.24] can be modelled electrically by the methods described in sections 6.3, 6.5 and 6.6. Since net long-wave flux:

$$L_N = L_o - L_i$$

it follows that:

$$L_N = \sigma T_0^4 - \sigma T_i^4$$

$$= \sigma(T_0^3 + T_0^2 T_i + T_0 T_i^2 + T_i^3)(T_0 - T_i) \qquad [6.25]$$

so that by referring to equation [6.20], the "effective net resistance" to the flux L_N, which may be regarded as responding to a temperature difference $(T_0 - T_i)$ is given by the term $1/\sigma(T_0^3 + T_0^2 T_i + T_0 T_i^2 + T_i^3)$. Modifying the example calculated in section 6.6 with $f_{(Gi)} = 1$ and $f_{(TV)} = 0.1$, the simulated "effective net radiative resistance" follows as:

$$R_N = \frac{1}{\sigma(T_0^3 + T_0^2 T_i + T_0 T_i^2 + T_i^3) f_{(Gi)} \cdot f_{(TV)}^{-1}} \qquad [6.26]$$

so that for $T_0 = 300°K$ and $T_i = 200°K$, $R_N = 266\ \Omega$. The temperatures would be simulated by potentials of 30 and 20 V respectively.

In order to allow for the energy flow involved in evaporation, an additional circuit similar to that built for observing the simulated sensible heat flux, H, should be made. The values chosen for the components depend on the Bowen ratio at the various height stages. Clearly, a Bowen ratio of 1 would require identical simulating circuits for both the H and LE fluxes.

Finally, the net short-wave radiation requires modelling. The most important property of this flux bears re-emphasis, in that it depends only on the incident short-wave radiation and on the surface albedo and not on any of the effective thermal resistances experienced by the other fluxes. Thus, if the net short-wave flux is to be simulated by an electrical current, the source resistance must be high compared to the total simulated flux resistance. In the example developed in this chapter, this would require a current source resistance in excess of 1000 Ω.

The entire micro-meteorological analogue circuit is shown in Fig. 6.5. Because of the need to simulate °K rather than °C, in order to deal with radiation, the potentials applied to both the terminations of the various circuit elements have to be considered carefully. By applying a half-wave rectified sinusoidal current to simulate the daily regime of net short-wave radiation, considerable realism can be introduced into the model. With a time scaling factor of $f_{(tT)} = 10^{-7}$ a basic frequency of $10^3/8.64 \approx 132\ \text{sec}^{-1}$ is required.

The most important parameter which can be observed in this experiment is the simulated surface temperature, particularly its diurnal amplitude and phase compared to the short-wave flux. Almost unlimited experimentation is possible, by virtue of the variations in simulated "effective thermal resistances" for both ground and atmospheric transfer processes. Albedo variations are simply simulated by controlling the simulating current amplitude.

A cathode ray oscilloscope should be used to monitor potentials and currents in the analogue circuits in order to deduce simulated temperatures and fluxes. Almost any laboratory oscilloscope is suitable for displaying potentials but the measurement of currents will in general require a two-probe differential input which should be connected across a known resistance in the path of the current to be monitored. A dual or multiple beam display will allow the observation of phase relationships between potentials at different levels.

A most important aspect of experimentation with this type of equipment is the simulation of the phase lag of the surface temperature with respect to the net short-wave radiation. In the absence of any energy loss by long-wave radiation, sensible atmospheric heating and evaporation (L_N, H and LE respectively), the phase lag would be exactly $\pi/4$ in accordance with heat conduction theory. However, the greater the sum: $(L_N + H + LE)$ compared with the ground flux G, the smaller the phase lag will be. The analogue circuits effectively allow the illustration of the phase and amplitude relationships existing between the fluxes of micro-meteorology as have been developed in the theory of climatonomy elsewhere (e.g. by H. Lettau, University of Wisconsin).

Although this type of experiment could be performed with a numerical computer, the analogue approach described here is more instructive as an introduction, having the advantage of observational realism.

The modelling of micro-meteorological fluxes allows accurate predictions to be made for these in defined environments when one or another flux is changed, either naturally or artificially. Natural changes include seasonal trends in the amplitude of the net solar flux and changes in the albedo of plant surfaces, the increase in ground conductivity (and capacity) after rainfall and the limitation of the evaporative flux by virtue of summer dryness. Artificial changes may be deliberate or inadvertent and include albedo modifications caused by the destruction or replacement of natural plant or other existing surfaces, some of which through the application of water-proof materials such as concrete or tarmac may almost completely eliminate the evaporative flux. One of the most readily sensed parameters in meteorology is the surface temperature, changes in which consequent upon any of the modifications or natural trends described above merit forecasting because of their importance to many forms of life.

References

The following texts provide further, general information on meteorological instruments and their use and may be of value (particularly Monteith, 1972) in assessing equipment of a wide variety of manufacture.

Baur, F. (Editor), 1953. *Linke's Meteorologisches Taschenbuch* (Neue Ausgabe). Geest and Portig, Leipzig.

Kaye, G.W.C. and Laby, T.H., 1959. *Tables of Physical and Chemical Constants*. Longmans, Green and Co., London, 12th edition.

List, R.J. (Editor), 1949. *Smithsonian Meteorological Tables*. Smithsonian Institution Press, Washington, D.C., 6th, revised edition.

Meteorological Office, 1963. *Pictorial Guide for the Maintenance of Meteorological Instruments*. H.M. Stationery Office, London.

Meteorological Office, 1969. *Handbook of Meteorological Instruments, 1. Instruments for Surface Observations*. H.M. Stationery Office, London.

Middleton, W.E. and Spilhaus, A.E., 1965. *Meteorological Instruments*. Toronto University Press, Toronto, Ont.

Monteith, J.L., 1972. *Survey of Instruments for Micro-Meteorology* (*I.B.P. Handbook No. 22*). Blackwell, Oxford.

The books listed below contain material of general reference:

Geiger, R., 1965. *The Climate near the Ground* (English translation). Harvard University Press, Cambridge, Mass.

Munn, R.E., 1966. *Descriptive Micro-Meteorology*. Academic Press, New York, N.Y.

Sellers, W.D., 1965. *Physical Climatology*. University of Chicago Press, Chicago, Ill.

The remaining references specifically add to information discussed in the chapters listed below:

Chapter 1

Hofmann, G., 1960. Meteorologisches Instrumentenpraktikum. *Universität München, Meteorologisches Institut Wissenschaftliche Mitteilung*, No. 7.

Chapter 2

I.G.Y. Instruction Manual, 4. Radiation Instruments and Measurements. Pergamon Press, New York, N.Y.

Kondratyev, K. Ya., 1969. *Radiation in the Atmosphere*. Academic Press, New York, N.Y.

Robinson, N., 1966. *Solar Radiation*. Elsevier, Amsterdam.

Chapter 3.

Byers, H.R., 1974. *General Meteorology*. McGraw-Hill, New York, N.Y., 4th edition, (chapters 4, 5 and 6).

Hess, S.L., 1959. *Introduction to Theoretical Meteorology*. Holt, Rinehart and Winston, New York, N.Y. (chapters 4, 6 and 7).

Any standard aerological diagram (e.g. Herlofsen skew T–$\log P$) will be found to constitute a concise summary of relevant information.

Chapter 4

Priestley, C.H.B., 1959. *Turbulent Transfer in the Lower Atmosphere*. University of Chicago Press, Chicago, Ill.

Sutton, O.G., 1953. *Micro-Meteorology*. McGraw-Hill, New York, N.Y.

Chapter 5

Lettau, H., 1954. Improved model of thermal diffusion in soil. *Transactions of the American Geophysical Union*, 35(1): 121.

Schwerdtfeger, P., 1970. *The Measurement of Heat Flow in the Ground and the Theory of Heat Flux Meters*. U.S.A. Cold Regions Research and Engineering Laboratory, Hanover, N.H., Report TR 232.

Chapter 6

Schwerdtfeger, P., 1964. An analogue computer for solving growth problems of floating ice. *Gerlands Beiträge zur Geophysik*, 73(1): 44–52.

Index